THEORY OF GRAVITY

GENERAL THEORY OF GRAVITY,
QUANTUM GRAVITY THEORY,
AND
HYDROGEN ORIGIN THEORY OF THE UNIVERSE

OTHER BOOK BY EFREN BASA ADUANA JR.

THEORY OF EVERYTHING

THEORY OF GRAVITY

GENERAL THEORY OF GRAVITY,
QUANTUM GRAVITY THEORY,
AND
HYDROGEN ORIGIN THEORY OF THE UNIVERSE

Efren Basa Aduana Jr.

THEORY OF GRAVITY: GENERAL THEORY OF GRAVITY, QUANTUM GRAVITY THEORY, AND HYDROGEN ORIGIN THEORY OF THE UNIVERSE

Copyright © 2016 by Efren Basa Aduana Jr.

All rights reserved. No part of this publication may be reproduced in any form or by any means electronic or mechanical, including photocopy, recording, or any information storage and retrieval system now known or to be invented, without permission in writing from the publisher, except by a reviewer who wishes to quote brief passages in connection with a review written for inclusion in a magazine, newspaper, broadcast, or on a website.

Edited and published by Efren Basa Aduana Jr.

ISBN-13: 978-1537186245
ISBN-10: 1537186248

In memory of our brother, M. Eric

Contents at a Glance

Preface xvii
Introduction xix

PART 1: THEORY ON GRAVITY
1 Cosmology and Existing Theories on Gravity 3
2 General Theory of Gravity 19
3 Quantum Gravity Theory 77
4 Theory of Everything: Standard Model with Gravity 99

PART 2: THEORY ON THE ORIGIN OF THE UNIVERSE
5 Cosmology and Theories on the Origin of the Universe 139
6 Hydrogen Origin Theory of the Universe 149

PART 3: OVERTHROWING THEORIES AND IDEAS
7 Overthrowing the Cosmic Background Radiation, the Observation of an Expanding and Accelerating Universe, and the Ideas of Dark Matter and Dark Energy 161
8 Overthrowing Einstein's Relativity Theories 171
9 Overthrowing Higgs Boson, String Theory, and the Question on the Dominance of Matter Over Antimatter in the Universe 207

Epilogue 219
Appendix A: Theory of Light 221
Appendix B: Flying Saucer 229
Appendix C: Overthrowing Charge and Parity 235
Notes and References 243
Bibliography 247
Acknowledgements 251
About the Author 253

Table of Contents

Preface xvii
Introduction xix

PART 1: THEORY ON GRAVITY

1   Cosmology and Existing Theories on Gravity 3
Isaac Newton: Law of Universal Gravitation 3
   What Was Newton Actually Thinking during Time He Thought
     of Gravity 8
René Descartes: Ether Plenum Vortex Theory 9
Albert Einstein: Theory of General Relativity 11
Standard Model and Gravity 12
Quantum Gravity Theory 15

2   General Theory of Gravity 19
Energy Field, Electromagnetic Field, Magnetic Field, and
   Gravitational Field 20
Dipole Magnet Energy Particle 22
Charge Theory: Energy Particle, Spin, and Energy Field 24
   Charge and the Direction of Emission of the Energy Field 26
   Bound and Free Particles 26
   Attraction and Repulsion 28
Why are the Equation of the Coulomb's Law and the
   Law of Universal Gravitation Similar to Each Other 29
Bond Angle and Gravity 34
General Theory of Gravity 36
   Polar gravity (Gravity-y): Repulsive and Attractive Force 37
   Field Spin Gravity (Gravity-x): Keeping or Holding Force 40
Newton's Mistake: Gravity on the Apple and the Moon 41

THEORY OF GRAVITY

Theory on the Elliptical Orbit of the Planets and Other Bodies 43
Theory on the Slow Spin of Venus 45
Is Gravity Weak Compared to Magnetism or Electromagnetism 49
Newton's Action-at-a-Distance 51
Spooky Action-at-a-Distance and Quantum Entanglement 52
    Einstein's Spooky Action-at-a-Distance 52
    Schrödinger's Quantum Entanglement 53
Time, Space, and Motion 55
Historical Development of the Concept of Inertia 57
    Aristotle 57
    Motion of the Planets and the Moon 57
    Nicolaus Copernicus 58
    Galileo 58
    René Descartes 60
    Isaac Newton 61
Mass 62
Weight 62
Disputation of Galileo's, Descartes', and Newton's Concept of Inertia 64
Inertia—New Definition 67
States of Inertia: A Body on the Surface of a Gravitational Producing Body 68
    A Body at Rest 68
    A Body in Motion 70
    A Body Thrown Straight Up from a Standstill 71
    A Body Thrown Straight Up From a Moving Vehicle 71
    A Body Dropped from a Height from a Standstill 73
    A Body Dropped from a Height While Moving 73
    A Body Inside a Ship that is Moving Up Very Fast 75
    A Body in a Free Fall Inside a Ship 75

3   Quantum Gravity Theory 77
Micro-scale/Micro-universe (Dipole Magnet Energy Particle): Quarks, Electron, Proton, Neutron, and Atom 78
    Definitions for the Dipole Magnet Energy Particle 78

Table of Contents

    Up Quark, Down Quark, and Electron 79
    Theory of How the Energy Field is Created by
        a Dipole Magnet Energy Particle 80
    Proton and Neutron 82
    Atom 83
Macro-scale/Macro-universe (Quantum Leap:
    Dipole Gravitational Body): Planet, Star, Solar System,
    and Galaxy 84
    Definitions for the Dipole Gravitational Bodies 84
    Moon, Planet, and Star 85
    Theory of How Gravity is Created by the Star, Planet, or Moon .. 86
    Theory of How the Spin of the Moon, Planet, Star, and Galaxy
        is Started 91
    Solar System and Planetary System 91
    Galaxy 92
Bode's Law: The Distance of the Planets from the Sun 93
Why are the Planets Located at their Distance from the Sun:
    Periodic Table of the Elements and the Gravitational Field 95
Quantum Gravity Theory in a Nutshell 96

4 Theory of Everything: Standard Model with Gravity 99

Fundamental Particles 101
    Dipole Magnet Energy Particle 101
    Charge/Spin (Charge Theory) 101
    Energy Field 102
    Mass (Mass Theory) 103
Fundamental Forces 103
    Strong Interaction 105
    Energy Field: Preliminary 105
    Quark Strong and Weak Force (Quark Theory):
        Strong Interaction 106
    Atom Strong and Weak Force (New Model of an Atom):
        Strong Nuclear Force 108
    Electromagnetism (Electricity) 112
    Light 114

Magnetism 115
Gravity 116
Nuclear Fusion 117
About Neutrino and the Source of the Neutrino......................121
Radioactivity (Formerly Weak Interaction) 122
On Maxwell's Electromagnetic Theory........................125
On Electroweak Theory 127
Summary: Energy Fields, Emissions, and the Unification of the Fundamental Forces 129
Standard Model—Revised 131
Theory of Everything in a Nutshell 134

PART 2: THEORY ON THE ORIGIN OF THE UNIVERSE

5 Cosmology and Theories on the Origin of the Universe **139**

Big Bang Theory, Steady-State Theory, and Einstein's Cosmology...140
Einstein's Cosmology 140
Creation of the Elements: The Hydrogen-Helium Abundances Issue 142
Steady-State Theory..143
Cosmic Microwave Background Radiation 144
Inflation Theory 144
Dark Matter, Dark Energy, and Einstein's Cosmological Constant 145
The Universe as Observed: Cosmic Background Radiation, Constant Temperature, and Filament 146
The Mistakes That Are Supporting the Big Bang Theory 147

6 Hydrogen Origin Theory of the Universe.................**149**

Hydrogen Origin Theory (HOT) of the Universe......................149
Early Nebulae 149
Star 149
Abundance of Hydrogen and Helium 150

Table of Contents

Galaxy 150
Supernova 150
Galaxy Clusters, Superclusters, and Filaments 151
Cosmic Background Radiation (CBR) 151
Constant Temperature of the Universe 152
Age of the Universe 152
Galaxy Classification 154
HOT Universe in a Nutshell 156

PART 3: OVERTHROWING THEORIES AND IDEAS

7 Overthrowing the Cosmic Background Radiation, Observation of an Expanding and Accelerating Universe, and the Idea of Dark Matter and Dark Energy 161

CBR as the Remnant of the Big Bang 161
Expanding and Accelerating Universe: Universe's Filament 162
    Expanding Universe 163
    Accelerating Universe 164
Dark Matter and Dark Energy 165
    Breaking Down the Dark Matter 166
    Dark Matter: Galaxy 166
    Dark Energy: Filament of the Universe 168
    Percentage of Dark Matter, Dark Energy, and Ordinary Matter . 170

8 Overthrowing Einstein's Relativity Theories 171

Ether 173
    Aristotle 173
    René Descartes 174
    Isaac Newton 174
    Christiaan Huygens 174
    Thomas Young 175
    Augustin-Jean Fresnel 175
    James Clerk Maxwell 175

THEORY OF GRAVITY

- Ether and Light 175
- Michelson and Morley Experiment 176
- FitzGerald and Lorentz: Contraction Hypothesis 176
- Theory of Special Relativity 177
- Theory of General Relativity.. 180
- How to Overthrow the Theories of Special Relativity and General Relativity... 182
- Inertia and Uniform Motion... 183
- Disputation of Galileo's Relativity... 183
- Disputation of Special Relativity (Constant Velocity)................. 185
  - Disputation of the Principle of Relativity............................... 185
  - Disputation of the Principle of the Constancy of the Speed of Light: Ether... 188
  - On Electromagnetism and Relativity...................................... 190
- Refuting the Effects of Special Relativity................................. 190
  - On Time Dilation 191
  - On Length Contraction 192
- Overthrowing Einstein's Understanding of Gravity.................... 192
  - Einstein's Understanding of Gravity..................................... 192
  - Overthrowing Einstein's Understanding of Gravity 193
- Disputation of General Relativity (Constant Acceleration)......... 194
  - Structure of Gravity in General Relativity............................. 195
  - Mechanism of Gravity in General Relativity 196
- Refuting the Proofs of General Relativity................................ 198
  - Gravitational Lensing 198
  - Orbit of Mercury 199
  - General Relativity's Spacetime Curvature and Elliptical Orbit of the Planets... 200
  - Black Hole 202
  - Gravitational Waves... 203
  - Gravitational Redshift 205
- Commentary 205

Table of Contents

9   Overthrowing Higgs Boson, String Theory, and the Question on the Dominance of Matter Over Antimatter in the Universe 207
Higgs Mechanism, Higgs Field, and Higgs Boson: Ether-Based
  Theory 207
  Higgs Field as Explained in Physics 207
  Higgs Field as Explained to a Layperson 209
  Higgs Mechanism as Depicted 209
  Higgs Discovery Announced and Higgs Awarded the
    Nobel Prize in Physics 210
  Higgs Field Explanation as the Source of Mass of the Particles . 211
  The Reality of Operating the LHC Regarding Jobs 211
  Limitations on the Use of the Particle Accelerator 211
  Refuting the Existence of the Higgs Boson 212
String Theory: Graviton 214
The Alliance of the Higgs Boson Theory and String Theory
  Theorists: Their Strategy and the Use of the
  Particle Accelerator 216
The Question of the Dominance of Matter over Antimatter
  in the Universe: Big Bang Theory 217

Epilogue 219
Appendix A: Theory of Light 221
Appendix B: Flying Saucer 229
Appendix C: Overthrowing Charge and Parity 235
Notes and References 243
Bibliography 247
Acknowledgments 251
About the Author 253

Preface

This book is the fulfilment of my three books on physics: *Theory of Light*, *Theory of Gravity*, and *Theory of Everything*. I started writing my first book on physics in the late 2009 and it took me nearly seven years, 2016, to finish all my three books. I thought before that I should write them in that order.

I had written my first book on physics, *Theory of Light*, but had not published it yet. I thought that the subject of light is not good enough to *garner* any interest and so I decided to write my next book. I had to decide which book is faster and easier to write: *Theory of Gravity* or *Theory of Everything*. I decided that it was the latter since the hunt for the Higgs boson was at its height. The search for the Higgs boson was then led by particle physics' prime machine, the Large Hadron Collider (LHC) of the European Organization for Nuclear Research (CERN) located beneath the border of Switzerland and France.

I finished my book *Theory of Everything* and self-published it in the late September 2013. I tried to do *something* about it but I decided not to pursue publicizing it. I thought that the attention it would get would affect me from writing my book on my theory of gravity.

After finishing my book *Theory of Everything*, I started going back to my early outlines for my book *Theory of Gravity*. (I had already outlined my theory of gravity during the time I was writing my theory of light. Partly, this outline was the reason why I decided to choose to work on my theory of everything instead.) I formally started looking into my theory of gravity around January 2014. I thought upon pondering on the subject of gravity that it would take me at least three years to solve its mysteries. Driven by some force, I work with a sense of urgency. I started reading, researching, and writing my theory on gravity in the late February 2014.

What I had learned later when I was writing my book, *Theory of Gravity*, is that those three books should have come out together, as one book need each other.

In my earlier readings on the subject of gravity, I came upon the subject of the search for the theory of quantum gravity. Having found a book titled *Three Roads to Quantum Gravity* by the physicist Lee Smolin published in 2001, I was struck by what I had read on page 211 where he intimated that: "We will have the basic framework of quantum gravity by 2010, 2015 at the outside." I thought that Smolin was being prophetic—until I learned later that the year 2015 is the centennial of Albert Einstein's publication of his theory of general relativity. (Wishful thinking on his part, perhaps?) I had written about general relativity before but it did not register in my mind that it would be 100 years in November 25, 2015. It is very timely that I wrote this book.

* * *

The field of particle physics and astrophysics is in a sad state right now as reality is thrown out of the window and that it is more about outrageous pronouncements, claims, and publications. The following are happening in physics right now:

- The Big Bang theory is being propped up even when there are many observed inconsistencies.
- The string theory is sucking the funding, time, and brainpower with no possibility of being tested.
- The discovery of the Higgs boson is one the biggest mistake in physics.

Preface

- The theory of general relativity is wrong and yet its proofs were purportedly discovered (the bending of light and gravitational waves).
- The "relativity" of the special relativity and general relativity based on Einstein's is wrong since his understanding of inertia was wrong.
- Nobel Prizes and other awards were given to theories and discoveries that were wrong.

The purpose of this book is to solve the problems of the understanding of gravity, to overthrow existing theories, and complete my theory of everything.

Partial theories and ideas that will be overthrown are:

- Higgs mechanism: Higgs boson (dubbed as the "God particle")
- String Theory and Graviton
- Big Bang Theory: Expansion and Acceleration of the Universe
- Cosmic Microwave Background Radiation
- Special Relativity: Einstein's Inertia and Relativity
- General Relativity: Bending of Light and Gravitational Waves
- Dark Matter and Dark Energy
- Charge Parity: Symmetry and Violation
- The question of the dominance of matter over anti-matter in the universe

My theories that will be used to overthrow the above theories are simple and coherent. The partial lists of my theories are:

- General Theory of Gravity
- Theory of Inertia
- Quantum Gravity Theory
- Theory of Everything

THEORY OF GRAVITY

- Charge Theory
- Mass Theory
- Quark Theory
- New Model of an Atom
- Standard Model with Gravity
- Theory of Light
- Hydrogen Origin Theory of the Universe

This book covers my theory of gravity and its related theories listed above. My theory of gravity is group into General Theory of Gravity, Theory of Inertia, Quantum Gravity Theory, and Hydrogen Origin Theory of the Universe. (With my Theory of Inertia or as it should be called the Law of Inertia, I had pulled the rug from under Albert Einstein's relativity theories.) My theory of everything is group into Theory of Everything, Standard Model (with Gravity), New Model of an Atom, Quark Theory, Mass Theory, Theory of Gravity, and Theory of Light. I had included my Theory of Light to complete my Theory of Everything and to show the end of the wave theory in quantum mechanics. My Theory of Light also sheds some idea on my unpublished book with the same title. I had also included the subject of Charge and Parity to show that my Charge Theory is right as it is based on observable reality and to overthrow the idea of Parity.

The strength of my theories is that they all agree with each other and they can explain all the aforementioned theories that I will overthrow. It will be noted in this book that in overthrowing the aforementioned theories I had presented a much better theory, exposed the flaw of the support of the contradicted theory, removed the proofs of the contradicted theory and explained it with my theory, or completely explained with my theory the claims of the contradicted theory. It is also for this reason that I had said in my book *Theory of Everything*, that there will be "no new Einstein"—because I am overthrowing his relativity theories.

Preface

Note that there are popular physicists that have criticized the current state of physics right now. Eric J. Lerner (*The Big Bang Never Happened*, 1992) and other scientists opposed the Big Bang theory (he espoused the plasma theory). Lee Smolin (*The Trouble with Physics: The Rise of String Theory, The Fall of Science, and What Comes Next*, 2007) and Peter Woit (*Not Even Wrong: The Failure of String Theory and the Search for Unity in the Physical Law*, 2007) questioned the dominance of string theory in the field of studies of physics and research. Alexander Unzicker (*The Higgs Fake: How the Particle Physicists Fooled the Nobel Committee*, 2013) questioned how the Higgs boson was discovered by the Large Hadron Collider (LHC). Jim Baggott (*Farewell to Reality: How Modern Physics Has Betrayed the Search for the Scientific Truth*, 2014) criticized the string theory and what is happening in physics right now.

The only way to correct physics right now is not only challenge its popular theories but also to offer theories that could solve the many problems in physics. I believed that my theories would be able finally to bring light into the world of physics.

Efren Basa Aduana Jr.
April 2016 [3.5]

Introduction

Those who will partake of the knowledge presented in this book will have their eyes opened to what is right and wrong in physics right now and realized that the empirical science is naked. As we search for truth, I open up with the sayings of these famous and influential physicists:

> Unthinking respect for authority is the enemy of truth.
> – Albert Einstein

> The greatest enemy of knowledge is not ignorance, it is the illusion of knowledge.
> – Stephen Hawking

This book is about overthrowing existing formidable theories that are supported by brilliant physicists and were supposedly tested and proven right. Having found this book, hopefully there should be an abundance of openness and honesty in reading my book.

These are the partial theories that will be corrected, explained, proven wrong, proven not to exist, or overthrown:

- Newton's law of universal gravitation
  - Explained. See *Chapter 2: General Theory of Gravity*.

THEORY OF GRAVITY

- Newton's explanation of gravity on the apple and the Moon
  - Corrected. See *Chapter 2: General Theory of Gravity* section on *Newton's Mistake: Gravity on the Apple and the Moon*.
- Principle of inertia: Galileo, Descartes, and Newton
  - Corrected. See *Chapter 2: General Theory of Gravity* section on *Disputation of Galileo's, Descartes', and Newton's Concept of Inertia*.
- Principle of inertia: Einstein
  - Overthrown. See *Chapter 2: General Theory of Gravity* section on *Inertia-New Definition*.
- Einstein's theory of special relativity
  - Overthrown. See *Chapter 2: General Theory of Gravity* section on *Inertia* and *Chapter 8: Overthrowing Einstein's Relativity Theories*.
- Einstein's theory of general relativity
  - Overthrown. See *Chapter 8: Overthrowing Einstein's Relativity Theories*.
- Black hole and wormhole
  - Overthrown. Proven not to exist. See *Chapter 8: Overthrowing Einstein's Relativity Theories*.
- Gravitational waves
  - **Awarded Special Breakthrough Prize 2016.** Overthrown. Observations relegated to the understanding of gravity. See *Chapter 8: Overthrowing Einstein's Relativity Theories*.
- Quantum gravity theory: unification of quantum mechanics and general relativity
  - Overthrown. See *Chapter 3: Quantum Gravity Theory* and *Chapter 8: Overthrowing Einstein's Relativity Theories*.
- Big Bang theory, cosmic microwave background radiation (CMBR) and its proofs
  - Overthrown. See *Chapter 6: Hydrogen Origin Theory of the Universe*.

Introduction

- Accelerating expansion of the universe
  - **Awarded Nobel Prize in Physics 2011.** Overthrown. See *Chapter 6: Hydrogen Origin Theory of the Universe.*
- Inflation theory relating to the Big Bang theory, which is attached to the standard model
  - Overthrown. See *Chapter 6: Hydrogen Origin Theory of the Universe.*
- Matter and antimatter symmetry as related to the Big Bang
  - Do not exist. See Big Bang theory above. See *Chapter 7: Overthrowing the Cosmic Background Radiation, the Observation of an Expanding and Accelerating Universe, and the Idea of Dark Matter and Dark Energy.*
- Dark matter and dark energy
  - Proven not to exist. Observation relegated to gravity. See *Chapter 2: General Theory of Gravity* and *Chapter 7: Overthrowing the Cosmic Background Radiation, the Observation of an Expanding and Accelerating Universe, and the Idea of Dark Matter and Dark Energy.*
- Graviton proposed in the standard model and as related to the string theory
  - Proven not to exist. See *Chapter 4: Theory of Everything: Standard Model with Gravity.*
- String theory
  - Proven wrong. There is no need for theory that cannot be tested. All of what it was touted to find, solve, or accomplish such as the gravity's graviton, quantum gravity theory, and theory of everything was already achieved by my aforementioned theories. See *Chapter 9: Overthrowing Higgs Boson, String Theory, and the Question on the Dominance of Matter over Antimatter in the Universe.*

- String theory's mathematically derived idea of extra dimensions, parallel universe, or multi-universe
  - Do not exist. See *Chapter 9: Overthrowing Higgs Boson, String Theory, and the Question on the Dominance of Matter over Antimatter in the Universe.*
- Higgs boson
  - **Awarded Nobel Prize in Physics 2013**. Overthrown. Higgs field, which is said to be a cosmic field, is another emanation of the idea of an ether. Mass is an intrinsic property of a moving energy particle, $E=mc^2$. The mass of a particle is related to its amount of energy and its speed. See *Chapter 9: Overthrowing Higgs Boson, String Theory, and the Question on the Dominance of Matter Over Antimatter in the Universe.*
- Charge-Parity concept
  - **Awarded Nobel Prize in Physics 1980**. Proven wrong. See *Appendix C: Overthrowing Charge and Parity.*
- Electroweak theory
  - **Awarded Nobel Prize in Physics 1979**. Overthrown. See *Chapter 4: Theory of Everything: Standard Model with Gravity.*

Organization of this Book

The book is organized into four parts. Part 1 is headed by Chapter 1, which is used as a background for Chapters 2, 3, and 4 to refute the existing ideas and theories on gravity. Part 2 is headed by Chapter 5, which is used as a background more on the current dominant theory of the origin of the universe (Big Bang theory) and the ideas, theories, and observations ascribed to it that will be refuted in Chapters 6 and 7. Part 3 is where in Chapter 8, Einstein's theories of special relativity and

Introduction

general relativity will be overthrown; and in Chapter 9, the dominant theories of the Higgs mechanism and string theory will be overthrown. Part 4 includes Appendix A, where I will forward a theory that light is only a particle and has no dual wave-particle property; in Appendix B I will show the application of my theory on gravity that will give us the idea of how to pursue gravity propulsion vehicle such as the flying saucer; and in Appendix C I will show that the idea of charge-parity is wrong.

Chapter 1: Cosmology and the Existing Theories on Gravity. Provides an overview of Newton's law of universal gravitation that will be corrected in Chapter 2; Descartes' ether vortex theory that will show where Einstein got his inspiration for his theory of gravity in his general relativity; Einstein's general relativity that will be progressively unraveled in Chapter 2, 3, 4, and finally in Chapter 8; the standard model and gravity, which what I had left in my book *Theory of Everything* that will finally be fulfilled in Chapter 2, 3, and finally Chapter 4; and quantum gravity that will be explained in Chapter 3 with a background in Chapter 2.

Chapter 2: General Theory of Gravity. Discuss the dipole magnet as the source of gravity, which explains the structure of the fundamental particles of the standard model; the correction of Newton's explanation of gravitation involving the apple and the Moon; the clarification of Newton's action-at-a-distance and Einstein's spooky action-at-a-distance; the discussions of the inertia of Aristotle, Galileo, Descartes, and Newton; the disputation of Galileo's, Descartes', and Newton's concept of inertia that will be used to overthrow Einstein's relativity theories; the correction of the concept of inertia; and the clarification of the different circumstances of the effect of inertia.

Chapter 3: Quantum Gravity Theory. Explains quantum gravity, which is a theory of gravity that can explain from the particles of matter to the bodies of the universe. This chapter is the continuation of Chapter 2.

Chapter 4: Theory of Everything: Standard Model with Gravity. Clarifies the fundamental forces I had started in my book *Theory of Everything* and finally includes gravity in the revised standard model. This chapter finally corrects the standard model that *unifies* the fundamental forces.

Chapter 5: Cosmology and Theories on the Origin of the Universe. Discuss the development of the theories of the origin of the universe and the observations on our universe that will be overthrown or explained in Chapter 6 and 7.

Chapter 6: Hydrogen Origin Theory of the Universe. Forwards a new theory of the origin of the universe that explains the current observations. This new theory will overthrow the Big Bang theory and ripple through other theories, ideas, and researches in physics.

Chapter 7: Overthrowing the Cosmic Background Radiation, the Observation of an Expanding and Accelerating Universe, and the Idea of Dark Matter and Dark Energy. Overthrows the idea that Cosmic Background Radiation is not the embers of the Big Bang. Since the Big Bang theory was refuted in Chapter 6, this chapter explains why we thought that the universe is expanding and accelerating. Explains what is dark matter and dark energy. (Dark matter and dark energy will not be found by CERN's Large Hadron Collider (LHC).)

Chapter 8: Overthrowing Einstein's Relativity Theories. Overthrows Einstein's special relativity and general relativity theories and shows what were the inspirations of Einstein's theories.

Chapter 9: Overthrowing Higgs Boson, String Theory, and the Question on the Dominance of Matter over Antimatter in the Universe. Refutes the existence of the Higgs boson, explains how the string theory is wrong, and explains the matter antimatter symmetry.

Appendix A: Theory of Light. Explains the structure (nature) of light as a fulfillment of what I had started in my book *Theory of Everything* and in Chapter 4 of this book. This is also a preview of my still unpublished book on my theory on light.

Appendix B: Flying Saucer. Shows what we can achieve with the right understanding of gravity that we cannot achieve with the erroneous general relativity.

Appendix C: Overthrowing Charge and Parity. About charge parity symmetry, charge parity violation, its relevance on the issue of the dominance of matter over antimatter as related to the Big Bang theory, and how the concept of charge and parity is wrong.

Convention on the Direction of Spin

To prevent confusion on the spin shown in the figures in this book, the figure below shows the convention on the depiction of the direction of spin of a particle (charge) and the spin of a gravitational producing body (counterclockwise and clockwise) looking in front or on top of a particle or on top of a gravitational producing body. The direction of the arrow points the direction of the travel of the free particle or the emission of the magnetic field of a particle or the gravitational field of a gravitational producing body. Positive (+) charge particle spins in a counterclockwise direction and negative (-) charge particle spins in a clockwise direction looking on top or in front of a particle. (Looking at the figures below, the "line" of travel that the particle makes or the spin of the axis of a gravitational producing body made is behind the direction of the spin.)

The convention becomes meaningful in the depiction of the direction of the spin of the fundamental particles particle (up quark, down quark, and electron) or the spin of the gravitational producing body and the mediating particle (photon) of the energy field (magnetic field, electromagnetic field, and gravitational field).

THEORY OF GRAVITY

The convention also becomes important in the depiction of the two particles with the same charge colliding head on with each other that result into repulsion.

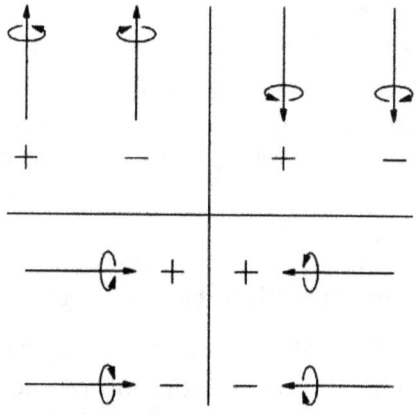

Part 1
Theory on Gravity

Chapter 1
Cosmology and Existing Theories on Gravity

Isaac Newton: Law of Universal Gravitation

Isaac Newton (1643-1727) was an English physicist, mathematician, astronomer, natural philosopher, alchemist, and theologian famous for his works in optics, mathematics, mechanics, and gravitation. (In mathematics, he developed differential and integral calculus independently with Leibniz. Gottfried Wilhelm von Leibniz (1646-1716) was a German mathematician and philosopher.)

In June 1661, Newton was admitted to Trinity College in Cambridge. In August 1665, his school closed as a precaution against the Great Plague so he went back to his home in Lincolnshire. He returned to Cambridge in 1667.

It was this break from school that in 1665 Newton conceived of the idea of gravity. Various accounts were written about how it actually happened. According to an account by John Conduitt, who was Newton's assistant at the Royal Mint and the husband of his niece, Newton was pensively meandering in the garden when he saw an apple fall from a tree.[1] According to another account by William Stukeley, a contemporary writer who wrote a memoir of his conversation with Newton in 1726, Newton recounted to him that the notion of gravity came to him during the time he was sitting in a contemplative mood when he saw an apple fall from a tree.[2] He thought of why the apple always fall perpendicular to the ground and not upward or sideways, that is, the apple always falls toward the center of the Earth.

Regardless of the varying accounts, Newton had an epiphany. He thought that the force that keeps the Moon in orbit is the same force that causes the apple to fall (Figure 1.1).

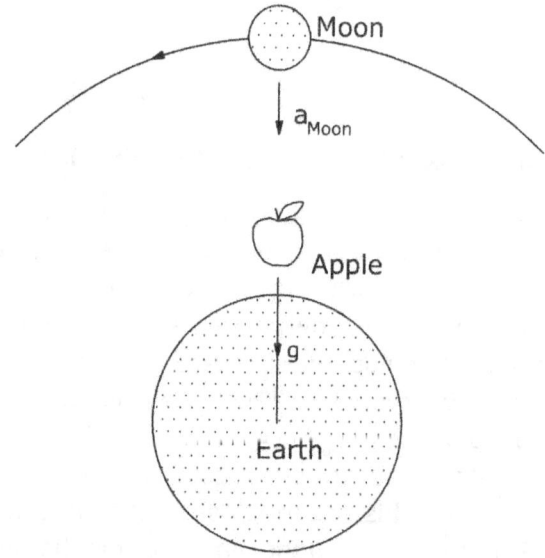

Fig. 1.1. Newton realized that the Moon is accelerating and its acceleration is caused by the centripetal force that he recognized as gravity, which is the same force that causes the apple to fall to the ground.

He recalled some thirty years later of the achievements he had made from 1665 to 1666 called his "anni mirabiles," his "miracle years":

Chapter 1: Cosmology and Existing Theories on Gravity

In the beginning of the year 1665 I found the Method of approximating series & the Rule for reducing any dignity of any Binomial into such a series. The same year in May I found the method of Tangents of Gregory & Slusius, & in November had the direct method of fluxions (as what Newton called the calculus) & the next year in January had the Theory of Colours & in May following I had entrance into ye inverse method of fluxions. And the same year I began to think of gravity extending to ye orb of the Moon & (having found out how to estimate the force with wch [a] globe revolving within a sphere presses the surface of the sphere) from Keplers rule of the periodical times of the Planets being in sesquialterate (one and a half times) proportion of their distances from the center of the Orbs, I deduced that the forces that keep the planets in their orbs must be reciprocally as the squares of their distances from the centers about which they revolve, and thereby compared to the force required to keep the Moon in her orb with the force of gravity at the surface of the Earth, and found them to answer pretty nearly. All this was in the two years of 1665-1666. For in those days I was in the prime of my age for invention & minded Mathematics & Philosophy more than at any time since.

(Newton was only twenty-three years old when he achieved all these feats. On the other hand, Einstein was 26 years old in 1905 when he published his theory of special relativity and 36 years old when he published his theory of general relativity, prompting Einstein to say that, "A person who has not made his great contribution to science before the age of 30 will never do so.")

Newton made the calculations comparing the ratio of the centripetal acceleration of the Moon to the acceleration of the apple on the Earth's surface with the ratio of the force on the Moon to that of the apple at the surface of the Earth. He used the period of the Moon to calculate for its centripetal acceleration. Also, he assumed that the gravitational force between two bodies is an inverse square of the distance between them (F ∝ $1/r^2$), known since 1640.

Newton was hesitant about publishing his work for the reasons that the value of the radius of the Earth was not accurately known at that time; he had only considered circular orbits, although Johannes Kepler had shown them to be elliptical; and that although he could not justify it at the time, he treated the Earth as if all its mass were concentrated at its center. (The proof for the latter was not completed until 1684.)

In 1679, Newton resumed his works on gravitation and its effects on the orbit of the planets as related to Kepler's laws of planetary motion. This was brought about by the brief exchange of letters with Robert Hooke (1635-1703). (Hooke at that year became the secretary of the Royal Society of London. Newton had earlier stayed away from the Royal Society when Hooke criticized some of his ideas. Hooke was at that time had been claiming that he had already discovered some of what Newton had thought about.) In a letter to Newton in 1680, Hooke raised some issues on whether it could be proven that an inverse square law leads to an elliptical orbit. Newton did not respond.

In 1684, Christopher Wren (1632-1723), the architect of St. Paul's Cathedral in London offered a prize for a proof that the inverse square law could lead to an elliptical orbit. Hooke claimed that he had the solution but would not divulge it, saying that—"in order that others may appreciate the difficulty involved." Naturally, Wren was unconvinced. In the summer of the same year, Edmund Halley (a young astronomer at that time who will become famous for his discovery of the Halley's Comet) went to Cambridge to ask Newton for the answer. Newton was said to have replied, "Why, an ellipse." Overjoyed, Halley ask Newton for the proof, of which Newton said he had misplaced it but promised Halley that he will redo his calculations and send it to him. The result of Newton's work was the publication of his full theory of gravitation in 1687, with

Chapter 1: Cosmology and Existing Theories on Gravity

Halley's financial help, in a book titled *Philosophiæ Naturalis Principia Mathematica* (Mathematical Principles of Natural Philosophy), now popularly known as *Principia*. (After the *Principia* was published, Hooke again claimed that he was the true discoverer of the law of gravitation, while Newton only filled-in the details. Newton was so contemptuous of Hooke that he did not publish his theory of light until after Hooke's death in 1703.) In this book, Newton used the Latin word *gravitas*, meaning weight or heaviness, for the effect that would become known as gravity.

* * *

Newton postulated that there is an attractive (gravitational) force for two point particles of masses m_1 and m_2 separated by a distance r. Newton's law of universal gravitation is given by the formula:

$$F = G \frac{m_1 m_2}{r^2}$$

where the value of gravitational constant is $G=6.67 \times 10^{-11}$ Nm^2/kg^2. (Newton did not have the value of G at that time.) Note that Newton's law of universal gravitation was stated only for point particles.

Newton's law of universal gravitation implied that two particles can interact directly with each other in free space. Newton called this phenomenon as "action at a distance." Newton's postulate of an invisible force that is able to act over vast distances led him to be criticized for introducing "occult" into science. Newton refuted this criticism in a letter to Richard Bentley:

> *That gravity should be innate inherent and essential to matter, so that one body may act upon another at a*

distance through a vacuum, without the mediation of anything else, by and through which their action or force may be conveyed from one to another, is to me so great an absurdity that I believe no man has in philosophical matters any competent faculty of thinking can ever fall into it. Gravity must be caused by an agent **acting constantly** according to certain laws, but whether this agent be material or immaterial is a question I have left to the consideration of my readers.

(Gravity is indeed innate to matter as its source is in the matter but only if a body has the structure and mechanism to produce gravity. We can see this from bodies with large mass such as a comet that does not produce gravity. This should be remembered later in the Einstein's theory of general relativity. Why the theory of universal gravitation is stated only for point particles will be explained in Chapter 2 on section *Why the Equation of the Coulomb's Law and the Law of Universal Gravitation are Similar to Each Other*. Gravity is explained in Chapter 2, 3, and 4.)

Some books would leave out the second sentence, which to me is very important in grasping Newton's understanding of gravity. Newton was actually describing the future understanding of gravity as a fundamental force with a mediating particle, the "agent," not known to him (and even until now). (It was not until the time of Michael Faraday when the existence of magnetic field was started to be acknowledged. Still, it is ironic that the right understanding of gravity was illusive even to the present time.)

What Was Newton Actually Thinking during the Time He Thought of Gravity

It may be thought impossible to know what Newton was actually thinking during the time he was "pensively meandering" in the garden or sitting in a "contemplative mood," absorbed in deep thought about gravity. We can infer that Newton was actually thinking about the inertia of Galileo and Descartes. (I had discussed inertia in Chapter 2 on *Historical Development of the Concept of Inertia*. Descartes' inertia practically follows that of Galileo's inertia.) Galileo explained his inertia that "a body will continue to move with a constant speed on a frictionless infinite horizontal plane," which is what the Moon does in orbiting around the Earth. Galileo realized that an external influence or force is needed to change the velocity of a moving object, not simply to maintain it, which is the centripetal force that Newton recognized as the gravity. Newton also thought of why the apple falls straight to the center of the Earth. Newton made some calculations and connected his two observations to come up with the idea of gravity.

Newton's formulation of his laws of motion attest to the refinements he did to the ideas of Galileo and Descartes on inertia.

René Descartes: Ether Plenum Vortex Theory

René Descartes (1596-1650) was a French philosopher, mathematician, and writer. His idea on gravity can be gleaned from his book *Principles of Philosophy* containing his laws of nature where he stated his idea of inertia and his idea of the universe.

Descartes also believed in the existence of the ether, which he thought of as a very dense medium of very small particles that pervaded all space. He believed that planets are pushed along their orbits by the vortex motion of the ether. In his solar vortex theory of gravity, he theorized that the Earth is located within an ether vortex with the Sun as the center (Figure 1.2). Descartes' universe was purely a mechanical universe.

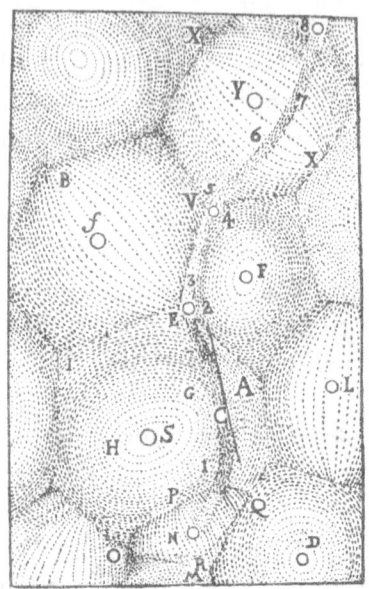

Fig. 1.2. Descartes' plenum vortex theory where the rotating motion of the Sun's ether vortex push the planets into orbit around the Sun. (Descartes' *Principle of Philosophy* Plate VI.)

(Descartes' plenum vortex theory is the spinning motion of the ether in space. Newton's bucket is a discussion on the motion regarding relative rotation and absolute rotation. Both of these may have contributed to Einstein's theory of general relativity, revising gravity from a force to a curvature of spacetime. As Einstein would say later: "We may assume the existence of an ether, only we must give up ascribing a definite state of motion to it."[3])

Albert Einstein: Theory of General Relativity

Albert Einstein (1879-1955) was a German-born American theoretical physicist who was famous for his theory of special relativity and theory of general relativity. He is regarded as the father of modern physics.

Chapter 1: Cosmology and Existing Theories on Gravity

Einstein graduated in 1900 with a diploma in mathematics and physics. Unable early on to find a teaching position, he got a job as a patent examiner at the patent office of Federal Office for Intellectual Property. However, Einstein still yearns to secure a job as a teacher. Einstein continued his work in physics, which culminated in his annus mirabilis of 1905 with the publication of his four papers on light quanta (photoelectric effect), size of molecules, existence of the atom (Brownian motion), and special relativity. His fifth paper pointed out the equivalence of matter and energy in his famous equation, $E=mc^2$ (where m is mass and c is the speed of light in a vacuum). From the $E=mc^2$, the theory of special relativity is said to tie mass and energy together. Special relativity predicts that nothing with mass can travel faster than the speed of light.

In 1907, Einstein thought that he could extend his relativity to the problem of gravity in what he called theory of general relativity. The reason Einstein formulated his general relativity was when he realized that his theory of special relativity (published in 1905) held that no interaction can propagate faster than the speed of light, which he said to be in contradiction to Newton's theory of gravity where gravity is a force that acts instantaneously between two distant objects (the so-called instantaneous action-at-a-distance). (From the above discussions on Newton, Newton never really said that the force of gravity acts *instantaneously* but rather he said that it acts *constantly*.) Also, that his special relativity applies only to constant velocity motion. Based from these arguments, Einstein decided to try to incorporate gravity into his relativity framework.

Einstein thought of the problem of general relativity while he was still working in the patent office in Bern. He had the realization that if a person falls freely, he will not feel his own weight. When he found the answer in 1907, he would later recall it as "the happiest thought" of his life. (It would take eight years of torment on him to finish his work.)

With Einstein's early papers, he garnered enough attention so that he got a job as a privatdozent (an unsalaried university lecturer) at the University of Bern in 1908. In 1909, he became a junior professor at the University of Zurich.

In 1915, Einstein published his theory of general relativity where he determined that massive objects caused a distortion (curvature) in the fabric of spacetime, which is felt as gravity.

Einstein gained his meteoric rise to popularity when Arthur Eddington (1882-1944), a British astrophysicist made an observation of a solar eclipse on May 29, 1919 that the light is bent by gravity proving Einstein's claim that the spacetime curvature created by a massive body can bend light.

In 1921, Einstein received his Nobel Prize in Physics for his work on the law of photoelectric effect.

Standard Model and Gravity

The standard model is a theory about the fundamental particles of matter and the fundamental forces of nature. The standard model is currently consists of twelve fundamental particles and three fundamental forces of electromagnetism, strong interaction (also called strong nuclear force or strong force), and weak interaction (also called weak nuclear force or weak force) with a notable absence of gravity (Figure 1.3).

(Note that the term strong nuclear force or strong force often mixed up with the term strong interaction will be cleared in Chapter 4.)

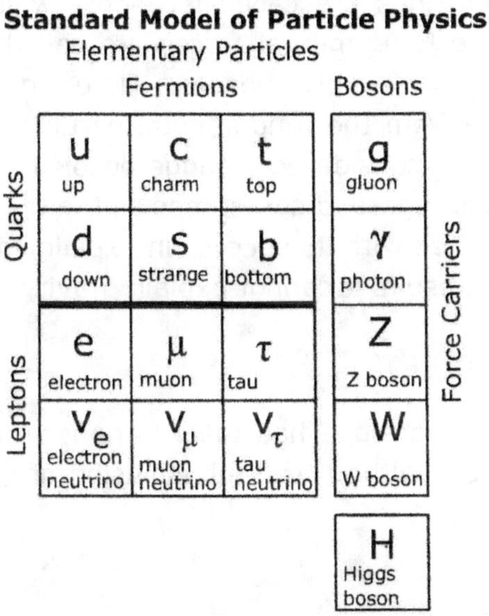

Chapter 1: Cosmology and Existing Theories on Gravity

Standard Model of Particle Physics
Elementary Particles

	Fermions			Bosons	
Quarks	u up	c charm	t top	g gluon	**Force Carriers**
	d down	s strange	b bottom	γ photon	
Leptons	e electron	μ muon	τ tau	Z Z boson	
	v_e electron neutrino	v_μ muon neutrino	v_τ tau neutrino	W W boson	
				H Higgs boson	

Fig. 1.3. Standard model of elementary particles with the inclusion of the Higgs boson.

The fundamental particles are the quarks: up, charm, and top, which have the charge of $+2e/3$; quarks: down, strange, and bottom, which have the charge of $-e/3$; leptons: electron, mu, and tau, which have a charge of $-e$; and neutrinos: electron neutrino, mu neutrino, and tau neutrino, which have no charge. The quarks commonly found in nature are the up quark and down quark. In the proton there are two up quarks and a down quark and in the neutron there are an up quark and two down quarks. The fundamental forces are mediated by the carrier particles of photon for light and electromagnetism, gluon for the strong nuclear force, and neutral Z^0 and charge $W^\pm$ for the weak nuclear force. The gluon of the strong nuclear force is the carrier particle of the quarks within the proton and neutron. The weak nuclear force refers to radioactive decay and nuclear fusion of subatomic particles. In particle physics terms, the quarks, leptons, and neutrinos are called fermions; while the mediating particles of the fundamental forces are called force carrier, gauge bosons, or simply bosons. Note that the fundamental forces came from the fundamental particles. (The so-called Higgs boson is not considered a gauge boson.)

The standard model is regarded sometimes as the theory of *almost* everything since even with its success in explaining wide varieties of experimental results it cannot explain gravity.

* * *

In my book *Theory of Everything,* I had made a revision of the current standard model but without the full inclusion of gravity (Figure 1.4).

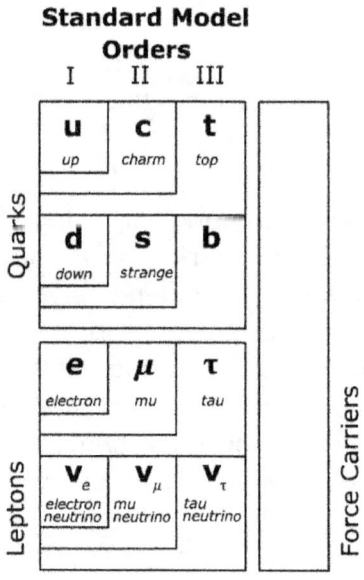

Fig. 1.4. Revised standard model from my book *Theory of Everything*.

The up, charm, and top quarks are understood to be one and the same particle, with the charm and top quarks as the

Chapter 1: Cosmology and Existing Theories on Gravity

higher order (or higher energy state) of the up quark. The down, strange, and bottom quarks are also one and the same particle, with the strange and bottom quarks as the higher order of the down quark. The electron, mu, and tau leptons are one and the same particle with the mu and tau as the higher order of the electron. The electron neutrino, mu neutrino, and tau neutrino are one and the same, with the mu neutrino and tau neutrino as the higher order of the electron neutrino. For now, the neutrino particles are still included in the fundamental particles but as they are the emissions from the fundamental particles, it will be understood that they belong to the fundamental forces. The fundamental forces are still left for now as they will be finally "unified" later in Chapter 4. (It is in this book not in my other book *Theory of Everything* that I separated electromagnetism into electricity and magnetism, and finally "unified" the fundamental forces by explaining the mediating particle of gravity.)

In this book, the knowledge of gravity will be finally integrated into the standard model. As it is often read about particle physics or astrophysics that the standard model cannot explain the dark matter and dark energy, both of them will be explained indirectly through the standard model (see *Chapter 3: Quantum Gravity Theory*). This book will also end the speculation that the string theory is the theory of everything that can solve the problem of gravity.

Quantum Gravity Theory

Quantum gravity is the term used for a theory that attempts to unify gravity of the theory of general relativity with quantum mechanics. It is said that the aim of quantum gravity is only to describe the quantum behavior of the gravitational field. On the other hand, the aim of theory of everything is the (mathematical) unification of all fundamental forces.

THEORY OF GRAVITY

Our current understanding of gravity is based on Einstein's theory of general relativity, which deals with the universe at the macroscopic scale. On the other hand, the fundamental forces (electromagnetism, strong force, and weak force) are described within the framework of quantum mechanics, which is in the microscopic scale.

Attempts to reconcile general relativity with quantum mechanics resulted in a theory that is not renormalizable, that is, the sum of all the forces do not cancel out and result in infinite value. (Renormalization is a procedure in quantum field theory where divergent parts of a calculation leading to nonsensical infinite results are absorbed by redefinition into a few measurable quantities yielding finite results.) Renormalization was first developed in quantum electrodynamics (QED) and was highly successful on electroweak theory, which incorporates the weak force with the electromagnetic force.

The problem with renormalization led to radical approach in solving quantum gravity through string theory and loop quantum gravity. String theory tries to unify gravity with the other fundamental forces. On the other hand, loop quantum gravity makes no attempt at unifying all the fundamental forces but instead tries to quantize gravitational field while keeping it separate from the other forces.

It is claimed that general relativity could be made compatible with quantum mechanics through the theoretical graviton, which is a spin-2 massless particle. The theoretical existence of graviton was derived from the observation that all fundamental forces except gravity have one or more known carrier particles. Theories such as the string theory, superstring theory, M-theory, and loop quantum gravity to some degree depend on the existence of graviton.

(It is ironic that although the graviton is vital to the search for the unification of quantum mechanics and general relativity it is not often heard of or written about. Instead, the discovery of the Higgs boson is hailed as the last remaining unknown

Chapter 1: Cosmology and Existing Theories on Gravity

particle of the standard model—a gross omission even by some physicists.)

Chapter 2
General Theory of Gravity

Note: The structure of this chapter is indicated by the three dashes.

Definition of Energy Field as Pertaining to Electromagnetic
   Field/Radiation, Magnetic Field, and Gravitational Field
Dipole Magnet Energy Particle
Charge Theory: Energy Particle, Spin, and Energy Field
Why the Equation of the Coulomb's Law and the
   Law of Universal Gravitation are Similar to Each Other
Bond Angle and Gravity
General Theory of Gravity
     Polar gravity (Gravity-y): Repulsive and Attractive Force
     Field Spin Gravity (Gravity-x): Keeping or Holding Force
---
Newton's Mistake: Gravity on the Apple and the Moon

Theory on the Elliptical Orbit of the Planets and Other Bodies
Theory on the Slow Spin of Venus
Is Gravity Weak Compared to Magnetism or Electromagnetism?

Newton's Action-at-a-Distance
Spooky Action-at-a-Distance and Quantum Entanglement
    Einstein's Spooky Action-at-a-Distance
    Schrödinger's Quantum Entanglement

THEORY OF GRAVITY

Time, Space, and Motion
Historical Development of the Concept of Inertia
    Aristotle
    Motion of the Planets and the Moon
    Nicolaus Copernicus
    Galileo
    René Descartes
    Isaac Newton
Mass
Weight
Disputation of Galileo's, Descartes', and Newton's Concept of Inertia
Inertia—New Definition
States of Inertia: A Body on the Surface of a Gravitational Producing Body
    A Body at Rest
    A Body in Motion
    A Body Thrown Straight Up from a Standstill
    A Body Thrown Straight Up From a Moving Vehicle
    A Body Dropped from a Height from Standstill
    A Body Dropped from a Height While Moving
    A Body Inside a Ship that is Moving Up Very Fast
    A Body in a Free Fall Inside a Ship

Energy Field: Electromagnetic Field, Magnetic Field, and Gravitational Field

Energy field is the general term that I used to mean the same for the following terms:

- Magnetic field: The emission of the subatomic particles (proton, neutron, and electron) inside the atom and the

quarks (up quark and down quark) inside the proton and neutron forming into an energy field.
- Electromagnetic field. The emission or radiation caused by the flow of the electrons in a conductor (electricity) forming into an energy field.
- Gravitational field. The emission of a gravitational producing body forming into an energy field. Note that the structure of the magnetic field of the magnet and the gravitational field are practically the same.

Electromagnetic field is the term that should be used specifically for electricity where the flow of electrons in a conductor produces the energy field. The term electromagnetic radiation should be defined as the emission of photons from the electromagnetic field such as in electricity and electronic communication.

There is still a confusion on the nature of light as it is still considered as an electromagnetic radiation called visible light within the range of the electromagnetic radiation spectrum (from the gamma rays to the radio waves), that is, light is considered as an electromagnetic waves. The light from the Sun (a star) came from the nuclear fusion that emits high-energy photons that travels in space. In *Appendix A: Theory of Light*, I will discuss and show that the photon of light does not emit an energy field unlike the electrons in a wire conductor. In *Chapter 4: Theory of Everything: Standard Model with Gravity*, I will separate light (the visible light range in the currently understood electromagnetic radiation spectrum, which is an emission of photon from various sources such as nuclear fusion from the star, electrical, or chemical) from electromagnetism. Electromagnetism will be separated into electricity and magnetism.

Magnetic field is the term that should be used specifically for an energy field created by the fundamental particles: electron and quarks, and the subatomic particles proton and neutron that are consist of up quark and down quark, and the atom that produces the magnet. That is, the magnetic field of the magnet is the proof of the energy field of the quarks and the atom.

Gravitational field is the gravity itself produced by the structure and mechanism of a gravitational producing body.

A compass (magnet) is affected by gravity and electromagnetic field. The Sun and Earth's gravitational field is also called magnetic field. The fact that the magnetic field, electromagnetic field, and gravitational field affect each other and that they can be called energy field shows us that there is an underlying commonality between them.

As we shall see later in *Chapter 4: Theory of Everything: Standard Model with Gravity*, the term energy field encompasses the fundamental forces of strong interaction, electromagnetism, and gravity. Strong interaction, strong nuclear force, and weak interaction will be considered as pertaining to where these energy fields are within the framework of the structure of the proton and neutron (using my Quark Theory) and the atom (using my New Model of an Atom). Weak interaction will be referred to as radioactivity but whether it will be considered as a kind of an energy field or a condition of an atom (particularly of a proton changing into a neutron) will be discussed in Chapter 4. The emitted charge particle photon of light is already under the fundamental force. The emitted neutral charge neutrino that is currently considered under the fundamental particle will be transferred under the fundamental force.

Dipole Magnet Energy Particle

The foundation of matter, which is of the whole (and only) universe, is the dipole magnet energy particle. A dipole magnet energy particle is an energy particle (presumed spherical in shape) that is through its spin produces its own energy field (Figure 2.1).

Chapter 2: General Theory of Gravity

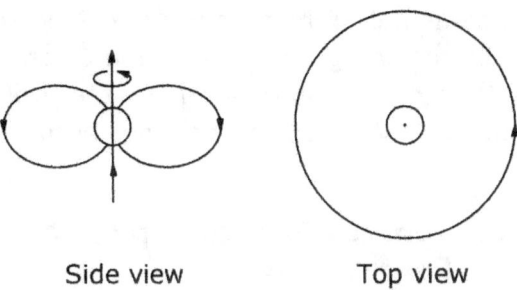

Side view   Top view

Fig. 2.1. The spinning dipole magnet energy particle and its energy field (magnetic field) shown on its side view and top view. (The spin shown is the direction of the spin of the particle. The direction of the spin of the particle determines its charge. This will be discussed under the subsection *Charge Theory* of the next section further below.)

The up quark, down quark, and electron, which are the only fundamental particles of matter, are dipole magnet energy particle. The energy field (magnetic field) of the dipole magnet energy particle has two components that are called polar magnetic field and field spin magnetic field (which is similar to the structure of the gravitational field, as we shall see later below in the section *Theory of Gravity*). The up quark governs the dipole magnet configuration of the proton, neutron, and the atom since it is the central particle comprising the nucleus of the said composite particles that is preserving the dipole magnet configuration, which is that of having the structure of the energy field of a dipole magnet energy particle.

All the energy fields of the fundamental particles and thus all fundamental forces of nature relating to these fundamental particles emanate from these dipole magnet energy particles—and nothing exists outside of them. (We can say that in the spirit of the atomist thinkers of ancient Greece, our universe is about the atom and the void.) That is, all bosons as the mediating particle of these fundamental forces came from these dipole magnet energy particles. (See the discussions in Chapter 4 on section *Fundamental Forces* regarding the revisions I had made on the fundamental forces and in *Chapter 9: Overthrowing Higgs Boson, String Theory, and the Question on the Dominance of Matter over Antimatter in the Universe* concerning the question on the source of the Higgs boson.)

Charge Theory: Energy Particle, Spin, and Energy Field

There are two kinds of particle: bound particle and free particle. Bound particle is within the atom (proton, neutron, and electron) or within the proton and neutron (up quark and down quark). Free particle is a particle that is traveling in space (photon of light, free proton, free electron, and neutrino) or inside a wire conductor in electricity (free electron).

In my book *Theory of Everything*, I had forwarded a Charge Theory that states that the spin and the direction of the spin (clockwise, counterclockwise, and not spinning) of a particle (and its energy field) is its charge, that is, spin and charge are the same (Figure 2.2).

The up quark, down quark, and electron are the only fundamental particles. The positive charge particle (up quark and proton) spins in a clockwise direction, negative charge particle (down quark and electron) spins in a counterclockwise direction.

Proton and neutron are composite particles composed of up quark and down quark. Proton is composed of two up quarks in the center and one down quark orbiting the up quarks. The resultant energy field of the quarks of the proton results in the energy field of the proton spinning in the counterclockwise direction (positive charge). Neutron is composed of one up quark in the center and two down quarks orbiting the up quark. The resultant energy field of the quarks of the neutron results in a non-spinning energy fields of the neutron. The explanation for the non-spinning energy field of the neutron and the

Chapter 2: General Theory of Gravity

spinning energy field of the proton is the reason for my support for the existence of the quarks.

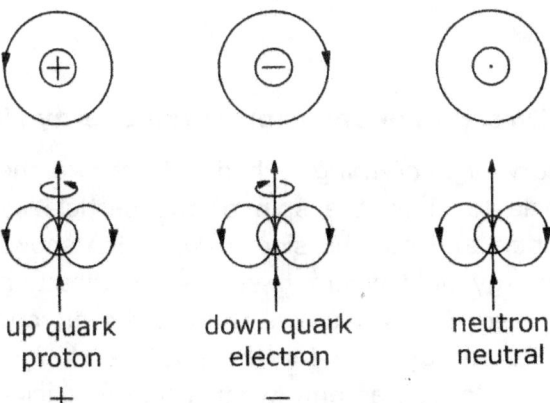

Fig. 2.2. Charge theory. Side view and top view of bound particles: Positive charge particle (up quark and proton) spins in a counter-clockwise direction, negative charge particle (down quark and electron) spins in a clockwise direction, and neutral charge particle (neutron) do not spin. Note that both the energy field of the positive and negative charge particles are emitted from the top and enters at the bottom.

A stable atom that is said to be "electrically neutral" is practically the same as the neutron, which has an energy field that does not spin. (The right term for "electrically neutral" should be "neutral charge.")

As I had explained in my book *Theory of Everything*, a fundamental particle can only be described as a "charge" particle as oppose to "charged" particle since it is its inherent property. On the other hand, an atom in its stable state is "neutral" (or currently described in physics as "electrically neutral") as it has an equal number of proton and electron but it could be described as "charged" by removing an electron to make it "positively charged" or by adding an electron to make it "negatively charged." (Based on my New Model of an Atom, considering that in a stable atom of an element there is an equal number of proton and neutron, I have not figured yet how an atom could take an extra electron or how it could have less electron.)

Charge and the Direction of Emission of the Energy Field

(In my book *Theory of Everything*, I had only shown the top view of the particle to show the spin of the particle as the charge of the particle and not the side view, which shows the structure of its energy field where it shows the direction the energy field is emitted. I was not sure then of the direction of the emission of the energy field for the positive charge and negative charge particle. It was only when I had to decide the issue of the idea of charge and parity discussed in *Appendix C: Charge and Parity* that I settled for the direction of emission of the energy field.)

Looking at the side view of bound particles (see Figure 2.2) the energy field is emitted from the top and enters at the bottom. The charge of the particle is determined looking from the top view of the particle. It is interesting to note that when the electron is emitted from the atom and is travelling in space (cathode ray tube) or within the wire conductor (electricity), the direction of emission of its energy field is also the direction of its travel, thus preserving its charge.

Bound and Free Particles

Particles could be classified as bound and free. Bound particles are particles such as the up and down quarks that are found inside the proton and neutron, and the proton, neutron, and electron found inside the atom (Figure 2.3a). Free particle such as the free electron flowing in a wire conductor create an energy field we called electromagnetic field (Figure 2.3b). Free particles such as the emitted photon, free proton, and free electron that are travelling outside of an atom or in space do not create an energy field (Figure 2.3c).

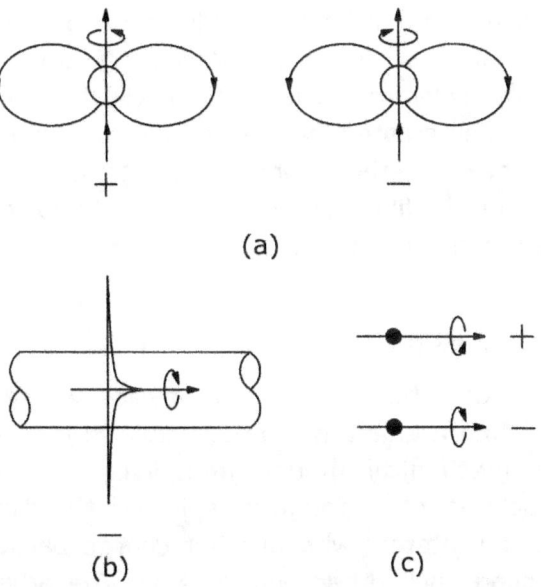

Fig. 2.3. The two kinds of particle are bound particle and free particle: (a) Bound particles with their energy field (b) Free electron inside a wire conductor create an energy field (electromagnetic field). (c) Photon, free electron, and free proton travel as a spinning particle with no energy field. (The spin shown is the direction of the spin of the particle.)

The photon and neutrino that travels in space do not emit an energy field. Photon is a negative charge particle (spins in a clockwise direction) and neutrino is a neutral charge particle (does not spin). The idea that a free particle could produce an energy field had bothered me for quite a while. This was an issue for the photon of light when I wrote my first book in physics on my theory of light where I thought that the structure similar to the one shown in Figure 2.3b explains its frequency and wavelength (see *Appendix A: Theory of Light*) but later I *decided* that the photon does not emit an energy field. Just as this is true for the photon of light, I believed that this is also true for the free electron and free proton in a particle accelerator or the electron in the cathode ray tube. For the electric current, which is the free electron flowing in a conductor wire, the electromagnetic field created by the electron may have been created by its energy field's interaction with the energy field of the atom. Think of the created electromagnetic field as like the wake of the ship or the shockwave created by a spinning bullet.

Attraction and Repulsion

Unlike the present definition of spin in physics where it is relegated only to being a quantum number, spin in my Charge Theory is a real mechanical motion and it is an inherent property of a particle as it is through its spin and the direction of its spin that it can interact with another charge particle by attracting or repelling each other (Figure 2.4). The attraction and repulsion of the particles makes our world solid.

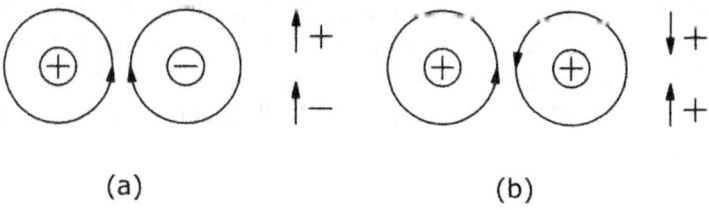

(a)   (b)

Fig. 2.4. Attraction and repulsion of charge particles according to my Charge Theory. Top view: (a) Opposite charge particles attract each other. (b) Same charge particles repel each other.

The interactions of charge particles could be between free particles (photon, free electron, and free proton), bound particles, or between bound and free particles (atom and photon). For this matter, a non-spinning neutrino can even go

Chapter 2: General Theory of Gravity

through the Earth without being affected because they do not spin and that the only way they could be affected or affect a particle or an atom is if they directly collide with a particle.

(It is known in physics that a neutrino has an antineutrino. To me this is clearly wrong since a neutrino has no charge, that is, it does not spin hence it has no antiparticle.)

Why are the Equation of the Coulomb's Law and the Law of Universal Gravitation Similar to Each Other

The force of gravitation between two bodies is given by the formula: $F=Gm_1m_2/r^2$, where G ($6.672 \times 10^{-11} Nm/kg^2$) is called gravitational constant, m_1 and m_2 are the mass of the two bodies, and r is the distance between the two bodies measured from their center (Figure 2.5).

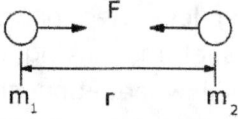

Fig. 2.5. Depiction of forces between two bodies.

The force between two charged particles is given by the formula: $F=kq_1q_2/d^2$, where k is 9×10^9 Nm/C^2, q_1 and q_2 are the charge of the two particles, and d is the distance between the two particles measured from their center (Figure 2.6).

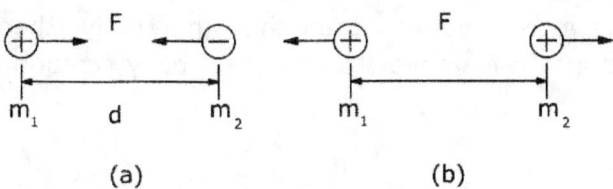

Fig. 2.6. Depiction of force between two charge particles. (a) Opposite charge attracts. (b) Same charge repels.

The symbol *k* is called Coulomb's law constant, which is said to be the value dependent upon the medium that the *charged particles* are immersed. (Note: There may be a confusion here as the Coulomb's law refers to "charged" particles, whereas I refer to a particle as a "charge" particle while a body is referred to as positively "charged" or negatively "charged" as I had discussed in the section *Charge Theory* above. Suffice to say, this understanding is still not yet settled in physics and so this section still carries this confusion.)

* * *

Coulomb's law is defined as a law of physics that describes the electrostatic interaction between electrically charged particles.

Judging from the definition of the Coulomb's law we know that the term "electrostatic" involves a body that is either positively charged with its atoms having less electrons or a body that is negatively charged with its atoms having more electrons. More so, while the Coulomb's law equation shows that of charge particles, it involves a "body" as the experiment uses a device that measured the force between the charge particles. In the experiment of the repulsion or attraction of charge particles, since this involves a "body" such as a ball or sphere, then this could be more or less a physical movement of the electron that created the repulsion or attraction. If both bodies have more electrons, they will repel each other as the electrons of both bodies will try to go to the opposite body but will be repelled instead—as if both bodies repel each other. If one body has more electrons than the other body, then the excess electrons will move to another body creating an attraction.

* * *

To understand the Coulomb's law is to understand that free particles (electron, proton, and photon) moving in space does not emit an energy field while bound particles (up quark, down quark, proton, and electron), that is, particles within the atom emit their own energy field. Thus, Figure 2.6 could only be a *representation* of the force between two bound charge particles derived from the experiment.

Based from the discussions above, the similarity between the equations of the Coulomb's law and the law of universal gravitation could be explained if they are both evaluated on a similar *set-up*.

The Coulomb's law makes sense if the bound charge particles have their own energy field that will result in either attraction or repulsion (Figure 2.7).

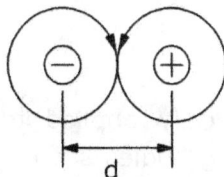

Fig. 2.7. Force between two opposite charge particles showing how through the spin of their energy field they attract each other.

In the interaction of the bound particles, the positive charge at the center is forcing the negative charge particle to orbit it such as the down quark orbiting the up quark and the electron orbiting a proton (Figure 2.8).

Note that the energy field of both particles emanates from their center.

(In the bound particles such as the up quark and down quark, it is possible that only the energy field of the up quark has the reach on the down quark. This is the same with the proton whose energy field has only the reach on the electron.)

THEORY OF GRAVITY

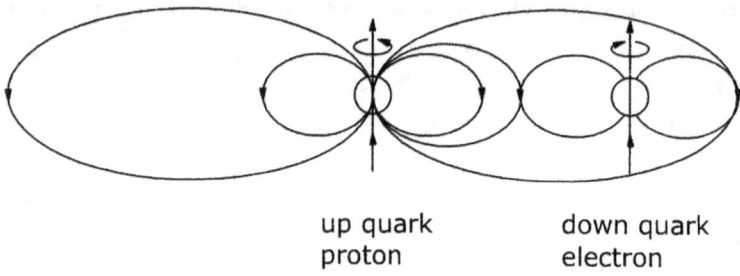

up quark   down quark
proton   electron

Fig. 2.8. Force between bound particles such as the up quark and the down quark or the proton and electron, where their energy field emanates from their center. (The spin shown is the direction of the spin of the particle.)

* * *

The law of universal gravitation is typically applied to the interaction between gravitational producing bodies such as the Sun and its planets and the planet and its moons (Figure 2.9).

Note that the body in the center is the one that is forcing another body to orbit it. The Sun, the planets, and the moons are gravitational producing bodies, that is, they produce their own gravitational field—practically like a giant magnet. We can see from Figure 2.9 that the gravitational field of gravitational producing body emanates from its axis. Thus, the force between the Sun and a planet or a planet and its moon is measured at their center since their gravitational field practically emanates from their axis.

(It is possible that only the gravitational field of the Sun is reaching the planet while the planet interacts with the Sun through its gravitational field that makes the Sun wobble. On the other hand, the effects of the Moon's gravitational field on the Earth could be observed with the changing of the tide.)

* * *

Chapter 2: General Theory of Gravity

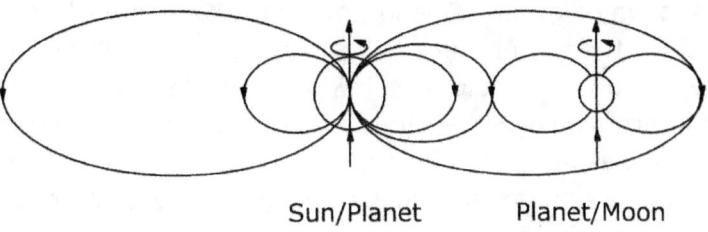

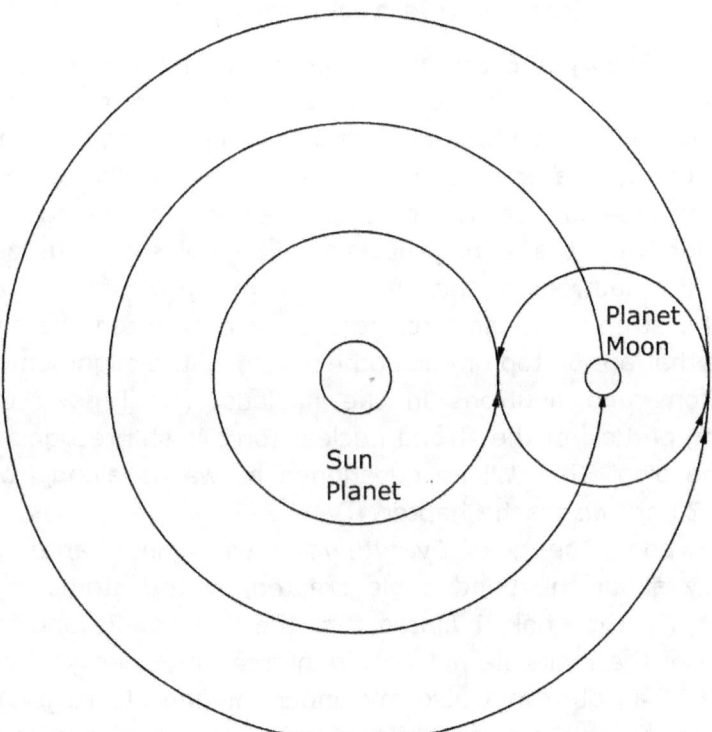

Fig. 2.9. The gravitational fields of the Sun and a planet or a planet and its moon where their gravitational field emanates from their center. (The spin shown is the direction of the spin of the Sun, planet, and moon is the spin of the body.)

We can see why Figures 2.7 and 2.8 for the Coulomb's law and Figure 2.9 for the law of universal gravitation are similar or that the equation of the Coulomb's law and the law of universal gravitation are similar since the particles and the gravitational producing bodies emit their energy field (magnetic field for the particles and gravitational field for the gravitational producing bodies) at their center.

Bond Angle and Gravity

In my book *Theory of Everything*, I had forwarded a New Model of an Atom where the structure of an atom is that of the proton and neutron aligned on top of each other in a pole and that only one electron orbit a proton. This is the reason why the protons in the nucleus do not fly apart as opposed to the current explanation that the strong nuclear force holds the lumped protons and neutrons in the nucleus. In my New Model of an Atom, the strong nuclear force is the force between the two particles that are on top of each other, such as the alignment of the protons and neutrons in the nucleus. (Until now, the mediating particle of the strong nuclear force is still recognized to be the pion. This will be questioned as we go along from Chapter 2 until we reach Chapter 4.)

I my book *Theory of Everything*, I wrote in *Chapter 7: Chemistry* about the bond angle created by the atoms in a molecule. (In the book, I *hinted* that the bond angle and the structure of the molecule not only reinforced my theory of the structure of an atom but also my understanding of gravity.) I had illustrated this concept using the molecules of ammonia (NH_3) and phosphine (PH_3) (Figure 2.10).

I specifically used as an example the ammonia and phosphine since they have the same number of hydrogen atoms and that any difference between the two can easily be observed. As I had said in my book, I actually know the gravity based from this example. What I meant was that as the phosphorous atom has more protons than the nitrogen atom, the phosphorous atom actually pulls together all the hydrogen atoms much closer. That is, the greater the number of protons of an atom, the stronger is its magnetic field. This knowledge is very much important in understanding why objects made of an element with more protons are heavy (weight) in the presence of gravity and why different parts of the Earth have different strengths of gravitation aside from the structure of its gravitational field (see Figure 2.20).

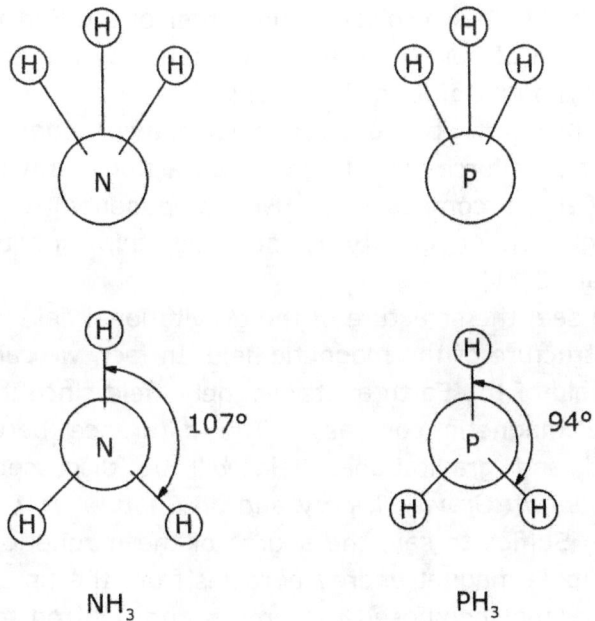

Fig. 2.10. The "ball and stick" model and structure formula of ammonia (NH_3) and phosphine (PH_3) and their bond angle.

General Theory of Gravity

Newton's gravity (law of universal gravitation) is a force of attraction while Einstein's gravity (theory of general relativity) is the curving of the fabric of spacetime due to mass. Einstein's gravity became the dominant idea of gravity in the present time. Most physicists try to explain the universe using Einstein's gravity but here on the surface of the Earth, the Moon, the escaping rocket ship, and the spaceships travelling in our solar system, Newton's gravity reign supreme. While we normally thought of gravity on Earth as a force that is everywhere and is pulling all the bodies down towards the center of the Earth and while we theorized of how the gravity is created by the Earth's core, we never thought of it as a mechanism.

The problem of gravity eludes us because we thought that gravity is just one force—the force of attraction. Gravity or gravitational field is composed of two components, which I called polar gravity or gravity-y and field spin gravity or gravity-x (Figure 2.11).

As we can see, the structure of the gravitational field is the same as the structure of the magnetic field. In fact, we call the gravitational field of the Earth as its magnetic field since it also influences the magnetic compass. (The difference between magnetic field and gravitational field will be discussed on *Chapter 3: Quantum Gravity Theory* and on *Chapter 4: Theory of Everything*.) Suffice to say, the source of the magnetic field is still in the dipole magnet energy particles from the up quark and down quark that composed the proton and neutron to the proton, neutron, and electron that composed the atom that is *scaled* up to the moon, planet, star, and galaxy.

(Note that mass has only correlation to the strength of gravity of a gravitational producing body in the sense that the bigger the mass, the more possible the larger and thus stronger the production of gravitational field. In refutation of general relativity, technically a large lump of material such as a comet can not necessarily produce gravity. This will be discussed in *Chapter 3: Quantum Gravity Theory*.)

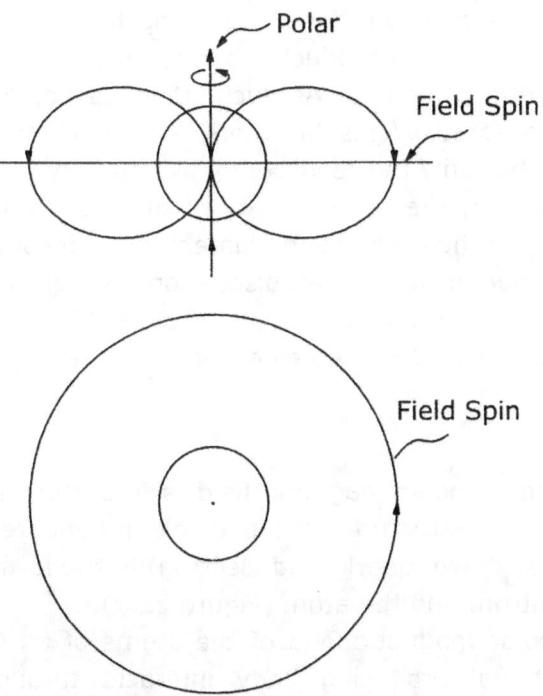

Fig. 2.11. The two components of gravity: polar gravity (gravity-y) and field spin gravity (gravity-x). (The spin shown is the direction of the spin of the gravitational producing body.)

Polar Gravity (Gravity-y): Repulsive and Attractive Force

There are two mechanism of polar gravity (gravity-y): one is from the poles of the large gravitational producing body and the other is the gravitational field of the large gravitational producing body acting on the polar magnetic field of the atom.

* * *

Polar gravity (gravity-y) is the north pole and the south pole gravitational field of the a gravitational producing body such as the moon, planet, star, and galaxy (see Figure 2.11). Just like a magnet, the poles of a gravitational producing body can repel or attract other gravitational producing body. (Proof of this is of course the magnetic compass.) As such, there is actually no such thing as "anti-gravity" just as there is no such thing as "anti-magnetism" but only the repulsion of like polarity.

For the most part, the polar gravity of the star in a solar system and galaxy in the cluster and filament is the most active factor in shaping our universe (see discussions of dark matter in *Chapter 7: Overthrowing the Idea of Dark Matter, Dark Energy, and the Expanding and Accelerating Universe*).

* * *

The understanding of polar magnetic field is important in the understanding of the structure of the dipole magnet energy particles (up quark, down, quark, and electron) in the formation of the proton, neutron, and the atom (Figure 2.12).

It is in the polar magnetic field of the atoms of an object that the gravitational producing body interacts through its gravitational field. Polar magnetic field is the quark strong force (what is understood currently as the strong interaction) and strong nuclear force of the fundamental forces of nature. The interaction of the polar magnetic field of a body and the gravitational field is what we called as inertia.

Chapter 2: General Theory of Gravity

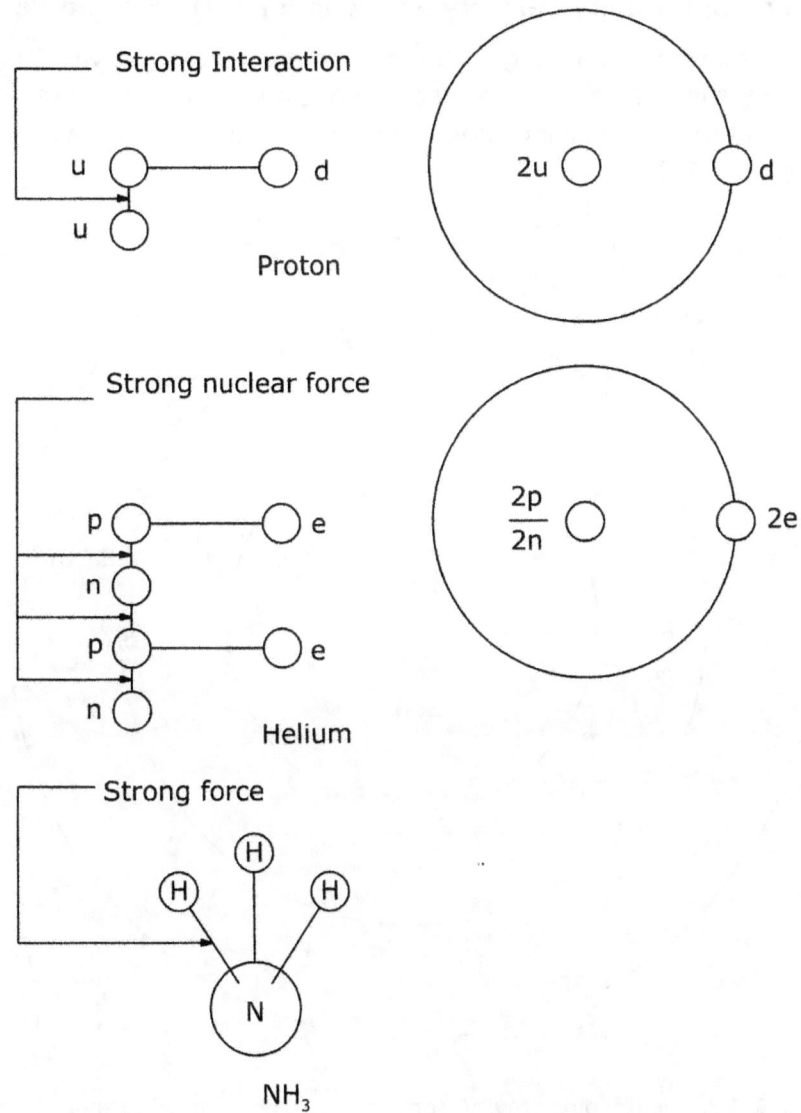

Fig. 2.12. Polar magnetic field is the strong interaction in the up quarks of proton and neutron, the strong nuclear force that binds the nucleus of an atom together, and the *strong force* that binds the molecule.

Field Spin Gravity (Gravity-x): Keeping or Holding Force

Field spin gravity is a gravitational component of gravity that makes the central gravitational producing body keep another gravitational producing body and make that body orbit it (Figure 2.13).

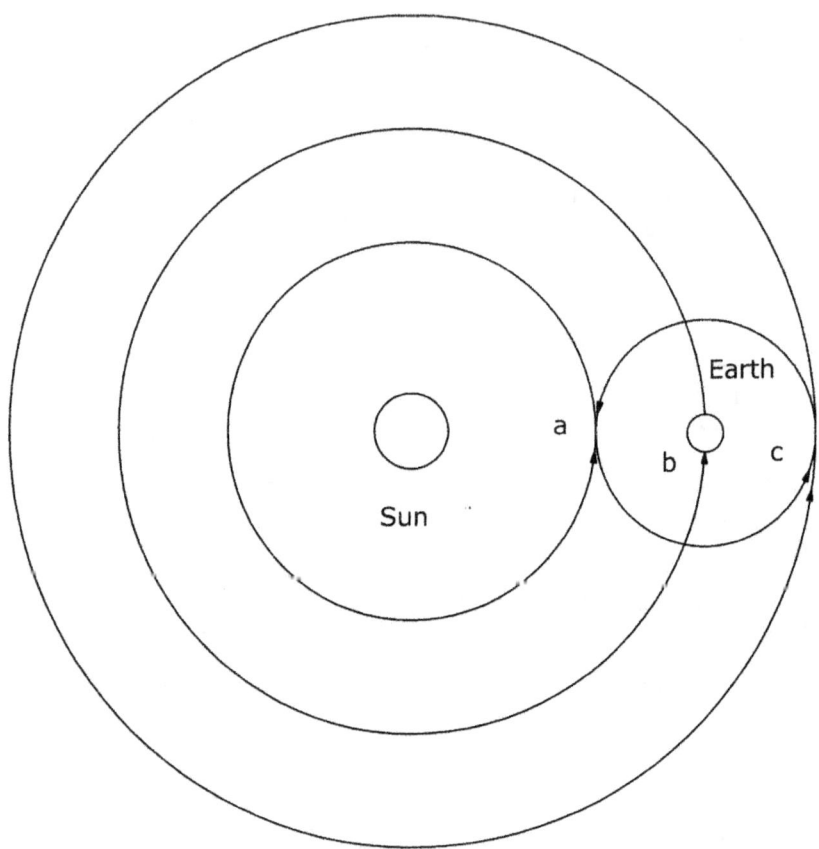

Fig. 2.13. Field spin gravity (gravity-x). The central gravitational producing body has stronger and far-reaching gravitational field than an orbiting gravitational producing body. At *a*, the gravitational producing bodies through their gravitational fields repel each other; at *b*, the gravitational producing bodies attract each other; and at *c*, the central body pushes the orbiting body in the direction it spins, which is in a counterclockwise direction.

Chapter 2: General Theory of Gravity

(Field spin gravity may explain the mechanics of the binary stars. Binary star is a system of two stars where one star orbits another star or both stars revolve around a common center.)

Note that this same explanation in Figure 2.13 can be used to explain how the up quark makes the down quark orbit it or the how the proton makes the electron orbit it. The immense speed by which the nucleus of a particle spins into orbit another particle forces it to its ecliptic plane. In the same way as the particle, a central gravitational producing body that makes another gravitational body orbit it forces (centrifugal force) it to its ecliptic plane.

The field spin gravity is the most active in the central body that makes another body orbit around it and so it governs the orbit of a moon around a planet, a planet around a star, or in the galaxy. (Elliptical and spiral galaxies should have massive star at its center or the dynamics of its material that produces a gravitational field.)

For the most part, the polar gravity of the star in a solar system and galaxy is the most active in shaping of our universe (see discussions dark matter in *Chapter 7: Overthrowing the Idea of Dark Matter, Dark Energy, and the Expanding and Accelerating Universe*).

Newton's Mistake: Gravity on the Apple and the Moon

Newton's understanding of gravity based on the falling apple and the Moon orbiting the Earth played a significant mistake in our understanding of gravity. Newton thought that the apple falls to the ground because of the force of gravity. He also thought that the Moon is accelerating due to centripetal force, which he thought of as the force of gravity. Newton made some rough calculations that he thought proved that the force acting on an apple is the same force that keeps the Moon in orbit around the Earth (see Chapter 1 on section *Isaac Newton: Law of Universal Gravitation*).

Newton's argument does not explain why the Moon is accelerating as it orbits the Earth. Newton had just accepted without any question the explanation of Galileo and Descartes regarding the motion of the Moon around the Earth (see section *Historical Development of the Concept of Inertia* further below). For Galileo, it was because it was the "natural" motion of the Moon around the Earth as it moves without friction *equidistant* to the Earth's center (actually, the orbit of the Moon is elliptical). For Descartes, it was because God is the primary cause of the Moon's motion.

As I had just discussed in my theory of gravity, the gravitational field of a gravitational producing body has two components called polar gravity and field spin gravity. However, on the surface of the Earth, bodies are just practically subject to the gravitational field, which we could think of simply as radiating from its center. (The gravitational field comes out from the North Pole and returns to the South Pole.) While it is the field spin gravity that makes the Moon orbit around the Earth, for the apple it is the gravitational field of the Earth that acts (attracts) on the polar magnetic field of the atoms of the apple, attracting it to the Earth (Figure 2.14). That is, technically, the force acting on the apple is not the same as the force acting on the Moon even though both are just the gravitational field of the Earth.

Since gravity's mediating particle is the same photon on both the polar gravity and the field spin gravity, then both will have the same speed—that is, the speed of photon, which is the speed of light (photon). Thus, naturally, Newton will get the same result when he calculated the speed of the acceleration of the Moon around the Earth to the speed of acceleration of the apple falling to the ground.

Chapter 2: General Theory of Gravity

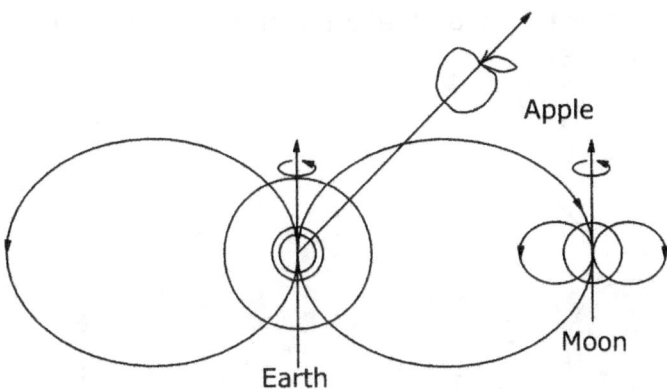

Fig. 2.14. The gravitational field of the Earth pierces through the poles (polar magnetic field) of the atoms of the apple while the Moon is forced into orbit by the Earth's gravitational field's field spin gravity component.

Theory on the Elliptical Orbit of the Planets and Other Bodies

The elliptical orbit of a planet around the Sun is often explained by how an ellipse is drawn, that is, with two focal points where one is the Sun and the other is an imaginary focal point. This is the case where we may be able to draw an elliptical path of the orbit of a planet but in reality, the elliptical shape of the path of a planet can be explained by some other means.

Johannes Kepler (1571-1630) determined purely by empirical means that the orbit of the planets around the Sun is elliptical. Isaac Newton (1643-1727) derived the theorem of elliptical orbit in *Book I, Section 9, Propositions 43-45* in his book *Principia*.

The understanding of the law of universal gravitation shown in the above sections on *Why the Equation of the Coulomb's Law and the Law of Universal Gravitation are Similar to Each Other* and *Theory of Gravity* discussed above showed that the field spin gravity is responsible for the orbit of the planets around the Sun or the orbit of the moon around a planet (Figure 2.15).

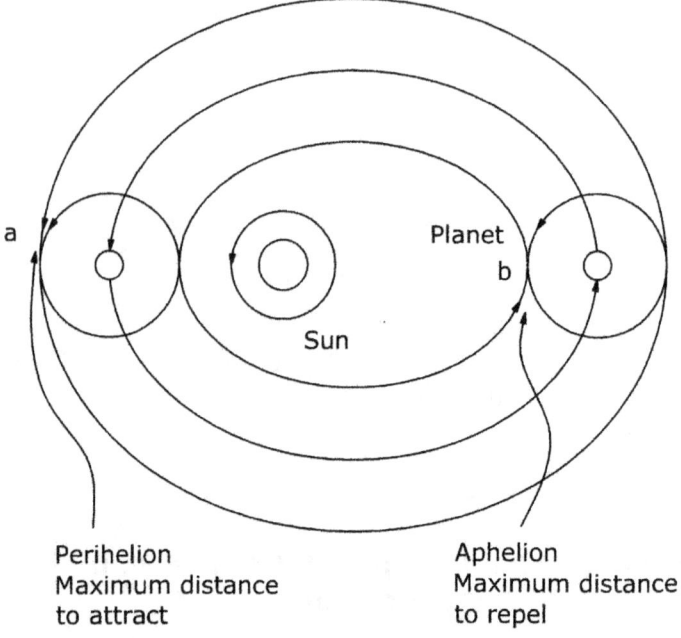

Perihelion
Maximum distance
to attract

Aphelion
Maximum distance
to repel

Fig. 2.15. Field spin gravity is responsible for the elliptical orbit of the planets round the Sun where the planet experienced the maximum attraction at point *a* and maximum repulsion at point *b*.

What Figure 2.15 showed is that the perihelion (closest distance from the Sun) is caused by the maximum attraction of the Sun to the planet while the aphelion (farthest distance from the Sun) is caused by the maximum repulsion of the Sun to the planet. Think of a planet orbiting at a distance of the perihelion from the Sun. The force of the attraction of the Sun will release the planet as if by a slingshot, called gravitational slingshot or gravity assist. (Even non-gravitational producing body such as an interplanetary spacecraft and probe uses gravitational slingshot.) However, as the planet reaches the aphelion, the Sun's gravity pulls the planet back again in orbit.

Chapter 2: General Theory of Gravity

Theory on the Slow Spin of Venus

When I wrote about the field spin gravity as the one responsible for the orbit of the planets around the Sun and the orbit of the Moon around the Earth, I thought of explaining what causes the slow spin of Venus. If my memory serves me, I remembered I had solved it a long time ago while I was writing my book *Theory of Everything*. I thought I had filed the answer away but I could not find it anymore and when I tried to remember how I did it, I could not even remember it anymore. After I was almost finished writing book, I thought of going back at the problem of explaining the slow spin of Venus since after all my theory of gravity should be able to explain it. Just as I learned from solving the problems and mysteries of science, truth can be rediscovered again.

* * *

It will be observed that the degree of tilt of Venus on its axis is that it is almost upside down or that it is spinning in a clockwise direction compared to other planets (Table 2.1, Tilt on Axis, 177°). One of the theories was that Venus was hit by a big body, forcing it to tilt that way. It will be observed also that Venus spins much slower compared to other planets and even compared to its length of orbit around the Sun (Table 2.1, Period of Rotation, 243d). It was observed that its atmosphere moves around the planet in four Earth days and so it was surmised that the atmosphere must be the cause of its slow spin or it contributed to its slow spin.

THEORY OF GRAVITY

Table 2.1 The tilt, length of revolution, and period of rotation of the planets.

Planet	Tilt on Axis	Length of Year*	Period of Rotation**
Mercury	**0°**	88d	**59d**
Venus	**177°**	224.7d	**243d**
Earth	23.5°	365.25d	23h 56m
Mars	25.2°	687d	24h 37m
Jupiter	3.1°	11.9y	9h 55m
Saturn	26.7°	29.5y	10h 40m
Uranus	**98°**	84y	17h
Neptune	28.8°	165y	16h 5m
Pluto	**120°**	247.7y	6d 9h

* In Earth years (y) or days (d).
** In Earth days (d), hours (h), minutes (m).

Observations:

1. The Sun spins (rotates) in a counterclockwise direction.
2. All the planets orbit the Sun in a counterclockwise direction regardless of the tilt of their axis.
3. Majority of the planets spins (rotates) in a counter-clockwise direction.
4. The length of orbit (Length of Year) of the planet around the Sun naturally follows the progression of their distance from the Sun.
5. At a tilt of 177°, Venus is almost upside down (180°) and is spinning (rotating) in a clockwise direction.
6. At a tilt of 98°, Uranus is practically rolling on its side.
7. At a tilt of 120°, Pluto is practically like Venus, which could be noticed from its long period of rotation in days instead of hours.
8. Mercury was observed to have a tilt of 0° but it can be seen that it has a very long period of rotation.

Chapter 2: General Theory of Gravity

Note: Owing to the current advances in science, we may have no reason to doubt the period of rotation of Mercury. However, as it can be observed from the periods of the planets that are spinning in a clockwise direction such as Venus and to a certain degree Pluto, astronomers should check if Mercury is instead rotating in a clockwise direction or what causes its lengthy period of rotation. Otherwise, we can also say that the strong gravitational field of the Sun acting on the material of Mercury like the Moon acting on the tide of the Earth, which is the polar magnetic field of the body, is the cause of the slow spin of Mercury.

* * *

As I have said, the field spin gravity is the one responsible for the orbit of the planets and should also be responsible for the slow spin of Venus in accordance to the tilt of Venus on its axis. Looking from the top view of the Sun and the planets, we can see that there is the repulsion and aiding (attraction) on the planets, which kept them from their position and there is the force from the Sun that keeps the planet into orbit (Figure 2.16).

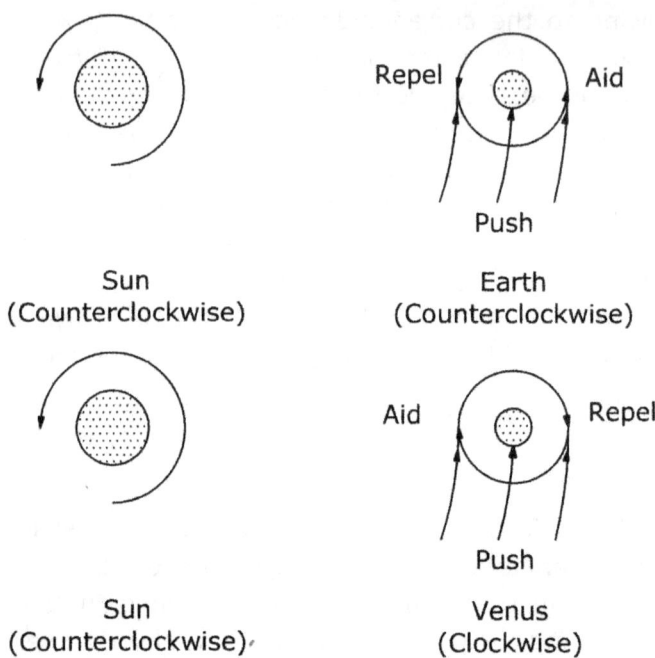

Fig. 2.16. Top view of the Sun and the planets (Earth and Venus) showing that field spin gravity acts practically the same for the two planets.

When I thought about what makes the spin of Venus slow compared to other planets, I thought of checking the side view of the gravitational field of the Sun and the planets as it affects the planets (Figure 2.17).

We can see from Figure 2.17 that for the planets such as the Earth, the gravitational field, the field spin gravity of the Sun aids or does not counter the gravitational field of the Earth. Looking at Venus, the gravitational field of Sun is actually being countered by the gravitational field of Venus, possibly slowing the spin of Venus.

Chapter 2: General Theory of Gravity

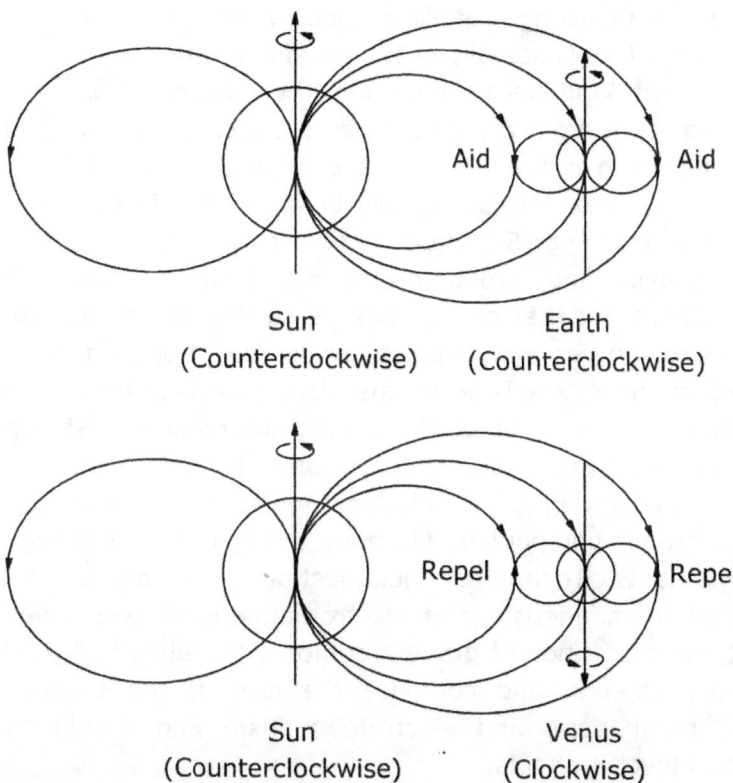

Fig. 2.17. Side view of the Sun and the planets (Earth and Venus) showing that the Sun's gravitational field aids the gravitational field of the Earth while it counters Venus' gravitational field. (The spin shown is the direction of the spin of the Sun, Venus, and Earth.)

Is Gravity Weak Compared to Magnetism or Electromagnetism?

It is said that gravity is weak in comparison to the fundamental forces of electromagnetism, weak interaction, and strong interaction. (I had separated electromagnetism into electromagnetism of electricity and magnetism of a magnet in Chapter 4.) The strength of gravity is often compared to the strength of a magnet in attracting metallic objects where the magnet wins over gravity. This is actually not a rational argument.

For gravity and weak interaction, there is no comparison at all, as gravity is from a gravitational producing body while the weak interaction is the process of changing of the neutron to proton. (The weak interaction will be discussed in *Chapter 4: Theory of Everything: Standard Model with Gravity*.)

For gravity and strong interaction, it will be understood later in Chapter 4 that gravity has its source within the atom. Gravity acts on *far away* objects while the strong interaction acts within the quark level or the strong nuclear force within the subatomic level. Naturally, strong interaction is stronger compared to gravity. Thus, we can only compare gravity to magnetism of a magnet and electromagnetism of electricity. To put into proper perspective, we have to organize our thoughts and arguments. To answer the question of "Is gravity weak compared to magnetism and electromagnetism?" we have to correct our perception of gravity, compare the ability of gravity to attract objects, and compare the size of the source of gravity, magnetism, and electromagnetism and its distance from the object it affects.

Our perception of gravity is that we do not really notice it until we make ourselves aware of it. On the surface of the Earth, we can feel gravity when we carry a heavy object, yet we seem to walk effortless, a leaf is blown in the wind, a bird fly in the air, and an airplane fly in the sky. An argument often cited for the weakness of gravity is that a magnet that attracts a paper clip wins over gravity. What we do not think is that even when the Earth is spinning on its axis we are on the surface of the Earth, things fall on the ground, planes needs power to fly, birds needs the air to fly, and a spaceship needs to overcome a certain escape velocity to be in outer space. The weight we experience is caused by the attraction of gravity. When we lie down in bed or rise from it, it is the gravity that is making us feel heavy but our body had compensated for it, and our mind and our muscles was already used to it. That is, our

Chapter 2: General Theory of Gravity

perception had changed because from a baby we struggled to train our muscle to overcome gravity and regain balance.

The comparison of gravity to magnetism and electromagnetism becomes apparent when we think that gravity attracts even nonmetallic objects while magnetism and electromagnetism can only attract metallic objects, especially those made of iron. We can say that it is the gravity that created our atmosphere and made our planets livable. It is also the Moon's gravity that attracts the bodies of water to create the tides.

Aside from our skewed perception of gravity, it is the size of the source and the distance from the source that is the source of the misunderstanding. From a very far distance, we can observe that the Sun attracts all the planets and the Earth attracts the Moon because of gravity. A magnet or an electromagnet cannot attract from a far distance due to their practical size. If we think about it, the source of the magnet is from the atoms. The source of electromagnetism is the flow of electrons that when configured into a coil becomes even stronger. On the other hand, gravity is created out of the dynamics within the gravitational producing body, making it enormous yet very far from the objects on the surface it affects.

(The practical argument about the strength of gravity, magnetism, and electromagnetism could lie in the means of the creation of the energy field (gravitational field, magnetic field, and electromagnetic field) or the energy of the photon particle. This is the subject in *Appendix B*.)

Newton's Action-at-a-Distance

Newton's action-at-a-distance was his understanding of gravity with the Earth's gravitational attraction of the Moon and the Sun's gravitational attraction of the planets. It was strictly an understanding of the great distance by which a large body attracts another body.

In his letter to Bentley (see Chapter 1 on section *Isaac Newton: Law of Universal Gravitation*), Newton explained his idea of "action-at-a-distance" of gravity. Newton believed that gravity, which is inherent to matter that enables one body to act upon another *at a distance* through a vacuum, is mediated by an "agent" that conveyed this action or force. Newton did not know if this "agent" is *material* or *immaterial* but he left it to the readers for consideration.

As I had explained above, we know now that "action-at-a-distance" is caused by the gravitational field and its "agent" is the mediating particle of the gravitational field (the same as the mediating particle of magnetic field), which is the photon. That is, the speed of gravity (photon of gravity) is the same as the speed of light (photon of light).

Spooky Action-at-a-Distance and Quantum Entanglement

In physics, the term spooky action-at-a-distance and quantum entanglement is often used to mean the same but they are actually different from each other. Einstein actually used the term spooky action-at-a-distance to refute quantum entanglement.

Einstein's Spooky Action-at-a-Distance

Spooky action-at-a-distance is the term that was also used by Einstein to refer to Newton's action-at-a-distance in the idea of an *instantaneous* action-at-a-distance. That is, Einstein was arguing on the idea that Newton's action-at-a-distance was operating at more than the speed of light, which Einstein's theory of special relativity prohibits. Newton did not actually refer to gravity's action-at-a-distance as acting *instantly* as it

Chapter 2: General Theory of Gravity

can be read from an excerpt of Newton's letter to Bentley. Rather, Newton referred to gravity as a *constant* action-at-a-distance. From this misunderstanding, the aim of Einstein's general relativity was to get rid of the spooky action-at-a-distance.

Schrödinger's Quantum Entanglement

Einstein's spooky action-at-a-distance is often erroneously thought to mean the same as the term quantum entanglement. And so, it seems that those who are not intimately familiar with the background of the argument on the terms thought that Einstein endorsed the idea of quantum entanglement when in fact it was the opposite. (Just like Fred Hoyle's coining of the term "big bang" to ridicule what later became called the Big Bang theory, Einstein was also "punished" to have the term "quantum entanglement" attributed to him.) To clear this misunderstanding, we have to go back to the understanding of Einstein and the arguments involving the two said terms.

Einstein thought of the incompleteness of quantum mechanics. He found a way to poke a hole at the armor of quantum mechanics in Heisenberg's uncertainty principle, which states that mechanical disturbance caused by the act of observation was a cause of uncertainty such that it is impossible to know both the precise position and momentum of a particle at the same time.

Einstein was bothered by the way quantum mechanics seemed to exhibit action-at-a-distance. In a thought experiment, Einstein posed a situation where two particles are set in motion with momentum and interacted with each other for a brief time. When the particles were already far apart, the observer measures the momentum of one of them. Then from the conditions of the experiment, the observer will be able to deduce the momentum of the other particle or if he measures the position of another particle, he will be able to know where the other particle is.

In 1935, Einstein together with Nathan Rosen and Boris Podolsky published a paper, which will become famously known as EPR paradox, reiterating his arguments on the behavior of the two particles that were "entangled" together. (Schrodinger coined the term "entanglement" to describe the correlations that exist between the two particles that interacted but are now distant from each other.)

During that time, Einstein had come to embrace the concept of realism and locality. In realism, he believed that a real factual situation exists independent of our observation. The concept of locality is that the two system that are spatially separated have an independent existence and that an action on one of these systems can not immediately affect the other. From this, we can say that Einstein do not believe in quantum entanglement.

The argument of quantum entanglement was actually a "Trojan horse" of Einstein to Bohr and Schrödinger who are the defenders of the quantum mechanics at that time. If the physicists had stayed with the idea that uncertainty principle applies only to within the atom, quantum mechanics would not be in quandary right now. Einstein was right in making his arguments against the problem he had given quantum mechanics of which its supporters mistakenly had taken in and had been taken in.

The reason why a large body such as the Sun influences the planet or a planet influences its moon is that their gravitational field extends very far due to the size of the source of their gravitational field. A free electron, which has much in common with the photon, is not the same as an electron within an atom, which is a dipole magnet energy particle with its energy field that is usually mostly confined only within the atom. That is, there is no way that the two particles that are very far apart from each other can "communicate" or even "know" the state of the other particle without the means of communicating through their energy field. If ever the experiment proved otherwise that indeed two different (energy) particles that was in contact before had now the same property is that the contact might have "even" out their different properties and not that they are "entangled" with each other.

Time, Space, and Motion

Newton's concept of space, time, dimension, and motion had been completely redefined by Einstein's relativity theories as Einstein was influenced by Ernst Mach's contradiction of Newton's concept of absolute time and space. Suffice to say, I advocated Newton's ideas.

Newton's *Scholium* in his book *Principia* defined time, space, place, and motion distinguishing them into absolute and relative, true and apparent, and mathematical and common. I had simplified my own definitions of time, space, and motion but also added position and dimension, which is practically under space. It will be understood in the definitions that I am setting up for the correct understanding of inertia from the definitions of absolute rest and absolute motion. (Absolute rest and relative motion in inertia are both the same.) Also, it will be understood from these definitions that my idea of our universe is that it is practically infinite and have unknown dimension that will be supported later by my theory in *Chapter 6: Hydrogen Origin Theory of the Universe*.

Absolute Space. A homogeneous and immovable space existing uniquely by itself. Bodies such as planets, solar system, galaxies, comets, nebulae, dust, hydrogen atoms, etc., are contained in this space.

Relative Position. A location of a point (or object) in relation to another point (or object) in absolute space.

Absolute Dimension. A conventional length of measurement use to measure absolute space or relative position.

Absolute Time. An immutable flow of time forward in one single stream.

Relative Time. A conventional fixed measurement of time based on a periodic counter used to mark an interval in the absolute time.

Absolute Motion. A body that is moving in absolute space or absolute dimension. Galaxies, nebulae, stars, solar system, planets cannot be surely known to move in absolute space except by observing their relative position. It is quite tricky to determine the absolute motion of our solar system or our galaxy but if we refer ourselves to a larger body in space then we can observe if we are moving or not moving relative to that body. A body moving in absolute dimension is in absolute motion. A person walking on land, a car running on the road, a plane flying in the sky, and a planet spinning on its axis and orbiting the Sun are in absolute motion as they cover a distance.

Absolute Rest. A body that is not moving in an absolute dimension. A body that is not moving on the surface of the Earth is in absolute rest. Note: Absolute rest is only meaningful to a body that is on the surface of a gravitational producing body. It is for this matter that our scientific instruments that rely on being completely still are able to work.

Relative Motion. The body that is not moving on the surface of the Earth is in relative motion as the Earth spins on its axis and is in orbit in the Sun. (In inertia, a body that is not moving inside a car, in a ship, or in a plane, although in relative motion to the Earth's surface, is actually in absolute motion.)

Chapter 2: General Theory of Gravity

Historical Development of the Concept of Inertia

The understanding of the concept of inertia is very much important as it is the lynch pin of Einstein's theory of special relativity and theory of general relativity. In trying to understand inertia, I went back to the sources of the concept of inertia starting from Galileo to Descartes and then to Newton, as Einstein had derived his knowledge of inertia from their works. (Einstein contributed a Foreword in the book *Galileo Galilei: Dialogue Concerning Two Chief World Systems* by Stephen Jay Gould, which "supposedly" contained Galileo's concept of inertia.)

Aristotle

Aristotle (335-322 BC), a Greek philosopher believed that the universe is composed of the four elements: earth, water, air, and fire, where each with its natural place one above the other. A natural motion is when an object moves vertically up or down towards its natural place. Heat rises above the air, bubbles rises above the water, and stone falls through air and water.

Any motion away from the natural place needed a cause: A cart does not move by itself; it has to be pulled by a horse. Based from his ideas, Aristotle believed that it is a natural tendency of an object to come to rest, as it is its "natural" state. Thus, objects on Earth are "naturally" at rest. An object that is moving is moved by something else. A force is required to keep the object moving and if the force is removed, then the object will come to rest.

Aristotle was primarily influenced by the accepted idea of his time that the Earth is the center of the universe.

Motion of the Planets and the Moon

Aristotle's ideas on motion appeals to common sense as normal observation shows that the natural state of an object is naturally at rest unless a force is applied to it. This is of course true for the objects on the Earth's surface where inclination (the level of the ground), air resistance, and surface friction are the factors that could make an object naturally stop from moving. What would baffle future philosophers was the continued motion of the bodies such as the Moon around the Earth and the planets around the Sun. From Galileo to Descartes to Newton, all these brilliant people will try to explain the motion of these bodies and apply it to the concept of inertia.

Nicolaus Copernicus

Nicolaus Copernicus (1473-1543) argued that the Earth and everything on it was in fact never "at rest," but is actually in constant motion around the Sun.[1]

Galileo

Galileo Galilei (1564-1642) an Italian physicist, mathematician, astronomer, and philosopher stated his principle of inertia (also sometimes referred to as the law of Inertia) in his *Dialogue Concerning Two New Sciences* published in 1637 as follows:

> *A body will continue to move with a constant speed on a frictionless infinite horizontal plane.*

So according to Galileo, inertia is the tendency of the body to maintain a constant velocity. Galileo thought that the moving object would only stop because of a force called friction. He also realized that external influence or force is needed to change the velocity of a moving object, not simply to maintain it.[2]

In an experiment using a v-shaped incline plane, he observed that a ball rolling down and then goes up the opposite side would reach *almost* the same height as the original height the ball was released (Figure 2.18).

Chapter 2: General Theory of Gravity

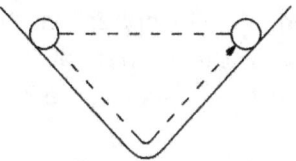

Fig. 2.18. A ball released on one side of an incline plane will reach the same height as the original height opposite of the incline even when one side of the incline is less steep.

Galileo reasoned that when the other side of the incline is made horizontal and friction is eliminated, the ball would roll with a constant speed along a straight line to infinity (Figure 2.19).

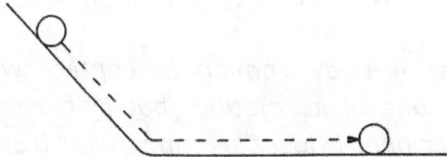

Fig. 2.19. When friction is eliminated and the second incline is horizontal, the ball will roll with a constant speed along a straight line and will never stop rolling.

However, Galileo's use of the term "horizontal" means equidistant from the center of the Earth, that is, when applied on a large scale, the ball rolling on a straight line will actually roll on a circular path.

Galileo also thought that the *circular motion* of the planet as it orbits the Sun or the Moon around the Earth was "natural" and therefore does not require any explanation.[3]

René Descartes

René Descartes (1596-1650), a French philosopher, mathematician, physicist, and writer is called the father of analytic geometry. Descartes' theory provided the basis for the calculus of Newton and Leibniz.

Descartes' *Principia Philosophiae* (Principles of Philosophy), published in 1644, was first intended to replace the Aristotelian textbooks that was then used in the universities of that time. In Descartes' *Principia*, he forwarded three laws of nature (Part II, Principles 37, 39, and 40):

> *First law of nature: A body always remains in the same state of rest; and once it is moved it continues moving. (Part II Pr. 37)*

> *Second law of nature: The movement of the body is along straight line and that bodies moving in a circle tend to move away from the center it is moving. (Part II Pr. 39)*

> *Third law of nature: A body coming in contact with a stronger one, loses none of its motion; but if it comes in contact with a weaker one, it loses as much as it transfers to that weaker body. (Part II Pr. 40)*

Note that Descartes' first law of nature stated that once a body is moving, it would continue moving. Why did Descartes have to include this statement? We can infer from this that it has to do with the movement of the planets around the Sun or the Moon around the Earth. Descartes provided the answer as to the initial source of this movement in Part II Principle 36:

> *That God is the primary cause of motion...*

Descartes' part of the second law of nature relating to the "movement of a body along a straight line" will become a part of Newton's first law of motion and another part relating to a

Chapter 2: General Theory of Gravity

body "moving in a circle tend to move away from the center it is moving," which is called the centrifugal force will become Newton's centripetal force that will become Newton's concept of gravitational force.

Descartes' third law of nature seems to be more only of his observation or his thoughts, as he tried to grasp the right understanding of the effect of a body hitting another body. Newton will be the one who will reach the correct understanding that will be stated in his third law of motion.

It was said that Newton named his book *Philosophiæ Naturalis Principia Mathematica* with the intention of displacing Descartes' *Principia* as Newton had always called his book *Principia*.

Isaac Newton

Isaac Newton (1643-1727) published his *Philosophiæ Naturalis Principia Mathematica* (famously known simply as *Principia*) in 1687. In this work he stated his three universal laws of motion, which at the present time is still very influential. Newton was able to synthesize and perfect his predecessor's ideas and added his own that showed his keen mastery of the subject. Newton's three laws of motion are:

> *First law of motion: Every body persists in its state of rest or uniform motion in a straight line, unless it is compelled to change that state by forces impressed upon it.*
> *Second law of motion: The net force acting on a particle of mass m produces acceleration a in the direction of the net force, $\sum F$. This law is expressed in the equation, $\sum F=ma$.*
>
> *Third law of motion (The law of action and reaction): For every action, there is an equal and opposite reaction.*

It will be observed that Newton had combined Descartes' first and second law of nature into his first law of motion but left the idea of a centrifugal force, which he will change later to centripetal force to explain the force of gravity. Newton had corrected Descartes' third law of nature into his third law of motion. As it will come down to us in the present time, Newton's first law of motion became the definition of inertia.

Newton's motion of the planets was relegated to gravity in stark contrast to Descartes' explanation that God is the source of the motion. Newton was very much aware of the lingering question on the motion of the planets.

> *Gravity explains the motions of the planets, but it cannot explain who <u>set</u> the planets in motion. God governs all things and knows all that is or can be done.*[4]

(What sets the motion of the planets will be discussed in Chapter 3 on subsection *Theory of How Spin is Started* under section *Macro-scale/Macro-universe: Dipole Gravitational Body): Planet, Star, Solar System, and Galaxy.*)

Mass

Mass is an inherent property of all fundamental particles of matter (up quark, down quark, and electron) derived from their amount of energy and their motion ($m=E/c^2$). The mass of a proton or neutron is composed of all their quarks (or the interaction of all their quarks). The mass of an atom is composed of all the protons, neutrons, and electrons (or the interaction of all their subatomic particles). The mass of a body is the total mass of all the atoms in it.

Weight

In physics, weight is defined as mass times the acceleration due to gravity, $W=mg$.

The mass of a body is the total mass of the atoms of a body. The atom of a body is a dipole magnet that is being attracted by the gravitational field of the gravitational producing body.

A body of a certain mass will have a different weight on Earth, on the Moon, or on another planet. The amount of the atom of the material and the kind of the atom of the element is a factor that determines the weight of a mass of a body on the surface of a gravitational producing body.

It was observed that objects fall with different gravitational acceleration on the different places of the Earth such as on the poles (g=9.832 m/s²), at 45° (g=9.806 m/s²), and on the equator (g=9.78 m/s²).[5] I believe that the reason for this is the structure of the gravitational field of the Earth. A body with a certain mass will weigh differently depending on the amount (volume) of gravitational field lines that a gravitational producing body produces and for the most part on the area of the gravitational producing body (Figure 2.20).

And so the question is, "What is *g* or the gravitational acceleration (or acceleration due to gravity)?" Gravitational acceleration or acceleration due to gravity could only mean the amount of gravitational field lines of a gravitational producing body. More gravitational field lines are coming out of the poles of a gravitational producing body than on its equator and so the poles have greater gravitational acceleration. Likewise, Earth is producing more gravitational field lines than the Moon for the Earth to have a much stronger gravitational acceleration than the Moon.

THEORY OF GRAVITY

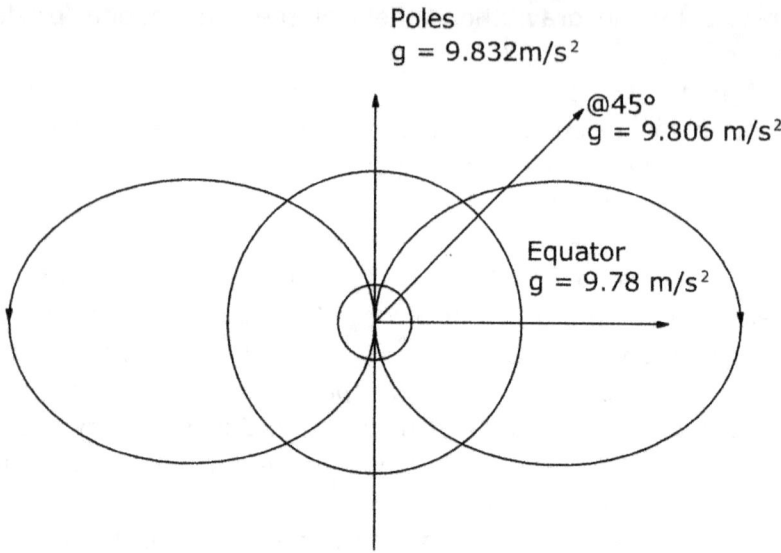

Fig. 2.20. The reason for the different gravitational acceleration on the surface of the Earth could be the structure of the Earth's gravitational field. There could be many gravitational field lines on the poles and decreasing towards the equator.

Disputation of Galileo's, Descartes', and Newton's Concept of Inertia

Our concept of inertia came down from Galileo to Descartes to Newton. Einstein's understanding of inertia was the basis of his theory of special relativity and theory of general relativity. Until now, our understanding of inertia had never been challenge or thought to be wrong. To lay a challenge on our understanding of inertia is to lay a challenge on Einstein's theories of relativity.

* * *

Galileo's idea of inertia is that a body will continue to move at a constant speed on a frictionless infinite horizontal plane, that is,

Chapter 2: General Theory of Gravity

equidistant from the center of the Earth. That is of course practically unrealistic as the only body that is moving continuously relative to Earth is the Moon, which according to Galileo needs no explanation as to why it orbits the Earth.

Descartes, on the other hand, in his first law of nature stated that a body will remain at rest but once it moves it continues moving. In his second law of nature, Descartes clarified that the movement of the body is along a straight line. We can say that in Descartes' laws of nature the concept of inertia is composed of two parts: one is that a body persists in its state of rest and the other is that once the body moves it will continue moving (along a straight line). In Descartes' first law of nature we can already see that the concept of inertia was slowly forming. Descartes' first law of nature is about a body that is at rest and the body that is in continuous motion. The latter refers more to the orbit of the Moon around the Earth and the planets around the Sun. Descartes had synthesized Galileo's idea but had kept Galileo's idea of a body that is continuously in orbit.

Newton's first law of motion, considered to be the law of inertia, combined, modified, and refined Descartes' first and second law of nature. Noticeably, it retained Descartes' description of a body that is "moving uniformly in a straight line," which could be traced back to Galileo. The discussion of his first law described air resistance and gravity as the forces that act on a body to stop it, reminiscent of Galileo's idea of friction.

As the understanding of inertia came down from Galileo to Descartes to Newton, nobody had really questioned the concept of inertia. Newton's first law of motion, which is the definition of inertia, states that:

> *Every body persists in its state of rest or uniform motion in a straight line, unless it is compelled to change that state by forces impressed upon it.*

THEORY OF GRAVITY

The definition of inertia can be divided into the following:

1. A body at rest persists in its existing state of rest, unless that state is changed by an external force.
2. A body in uniform motion in a straight line persists in its existing state of uniform motion, unless that state is changed by an external force.

The term "body at rest" is referred to a body that it is on the surface of the Earth and that it is not a gravitational producing body. The "body at rest" also means that the body is at rest but then it will be in motion if a force is acted on it. That is, the body at rest is at rest because there is the force of gravity acting on it not to move and then the body that is in motion will later stop due to gravity (weight), surface friction, brake, and air resistance.

The term "body in uniform motion in a straight line" could only mean that it is a gravitational producing body such as the Moon that is in orbit around the Earth (although its orbit could not be in equidistant to the Earth's center as Galileo had thought about as planets have elliptical orbits) since non-gravitational producing body such as a satellite or space debris could either fly away in outer space or fall back to Earth later. That is, the body is under the force of the field spin gravity. The "body in uniform motion in a straight line" such as the Moon orbiting the Earth, after having been under the influence of the Earth's gravitational field (the force that is acting on the Moon) will be in motion orbiting the Earth practically indefinitely.

Gravitational producing body acts on a non-gravitational producing body by attracting it, thus the former resisting the movement of the latter. On the other hand, a bigger and stronger gravitational producing body acts on another gravitational producing body, such as the Earth on the Moon and the planets on the Sun, by making the latter orbit the former practically forever. This is the big contradiction on the parts of the definition of inertia. This contradiction practically brings down Newton's first law of motion, which is the definition of inertia. To settle this contradiction, we have to go back to the intention of why inertia was thought of. Inertia was really thought of in the beginning as a property of matter to resists its movement. That is, our concept of inertia had been wrong all along. The solution to this dilemma is either to discard the term inertia or to redefine it for the purpose only of describing this behavior of a body or the effect of gravity on a body.

Inertia—New Definition

(Note that the original definition of inertia does not involve gravity at all but rather accord it as a property of matter, although it can be argued that it is implied.)

A body (matter) is composed of an atom, which is a dipole magnet particle. The mass of a body is the mass of all the atoms in a body. Weight is the effect of gravity attracting the atoms of the body, that is, the gravitational field attracting the dipole magnet particle atom. Thus, inertia is the attraction of the body by gravity. By taking the present understanding of inertia in the viewpoint of a body, inertia is defined as:

A body in the presence of gravity resists its movement.

It is understandable from the above definition that it is about the body since the laws of motion are about a body but in the viewpoint of gravity, inertia is defined as:

Gravity resists the movement of a body.

A body in this sense is a non-gravitational producing body.
The non-gravitational producing body in the presence of gravity whether at rest (absolute rest) or in motion (absolute motion) is under inertia as gravity will force the body to stop its motion. Absolute rest and relative motion are the same for a

body on the surface of a gravitational producing body. The body in motion could be moving across or moving up or down along the gravitational field lines. For simplicity, the gravitational field lines are viewed as emanating from the center of the gravitational producing body and projecting outwards (see Figures 2.22 and 2.23).

States of Inertia: A Non-Gravitational Producing Body on the Surface of a Gravitational Producing Body

A body on the surface of a gravitational producing body experience inertia, whether it is moving across, going up, or going down the gravitational field lines. The gravitational field lines (the gravity itself with no identification of being a polar gravity or field spin gravity) is viewed as coming from the center and going outwards.

A very good demonstration of inertia is that of a table covered with a tablecloth and on top of the tablecloth are a dish and a glass (Figure 2.21).

A Body at Rest

A body at rest (absolute rest) on the surface of the Earth (in the presence of gravity) will resists any change in its state unless a change in its inclination makes it move by itself or a force strong enough to move it acts on it.

An object that is in the state of absolute rest on a level surface of the Earth moves with the spin of the Earth (Figure 2.22).

The gravitational field of the Earth passes through every atom its material (from its source of gravitational field to its surface) and then passes through the material of the body on its surface to attract that body. The gravitational field attracting the atoms of the body experienced as the weight of the body

Chapter 2: General Theory of Gravity

resists from being moved even as the Earth is spinning on its axis. (The Earth spins at about 1,000 miles per hour on the equator.)

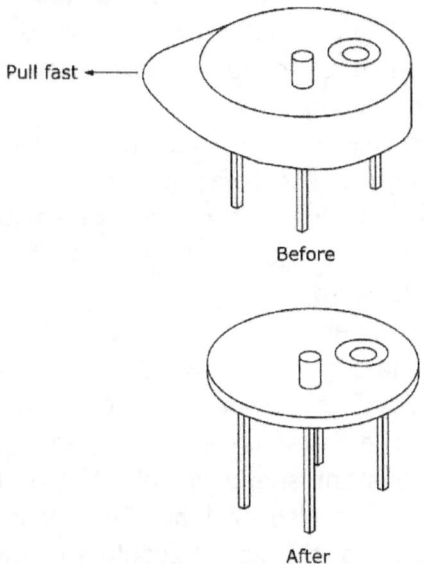

Fig. 2.21. A "magic trick" involving inertia. Pulling the tablecloth very fast and the dish and glass still stays on top of the table.

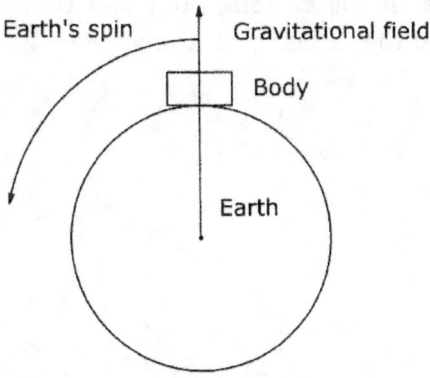

Fig. 2.22. A body at rest on the surface of the Earth moves with the spin of the Earth. The body is in a state of absolute rest.

THEORY OF GRAVITY

A Body in Motion

A body in motion (absolute motion) on the surface of the Earth (in the presence of gravity) is under inertia, that is, the body is under the attraction of gravity (Figure 2.23). A body in motion will experience surface friction due to its weight and air resistance that will make it stop moving.

A driver who suddenly accelerates his car feels being pushed back to his seat. A jetfighter pilot that accelerates will experienced being pushed back to his seat. These are similar to the example of the effect of inertia on the table with a tablecloth and dishes and glasses on top of it.

Traveling at a constant speed seems to present a problem to the idea of inertia. (This is the argument that Einstein used for his special theory of relativity.) It can be observed that a glass of water is not spilled or a person can walk normally inside a plane travelling at a constant speed and so it is as if that glass of water or that person is also on land. The reason for this is that the change of speed of the vehicle could spill the water in the glass but that the person can adjust to sudden changes in speed.

To refute Einstein's inertia, an experiment can be done by having a level surface where a metal ball is on top of a vehicle moving in a uniform motion. It will be found out that the metal ball should move opposite to the direction of the vehicle.

Chapter 2: General Theory of Gravity

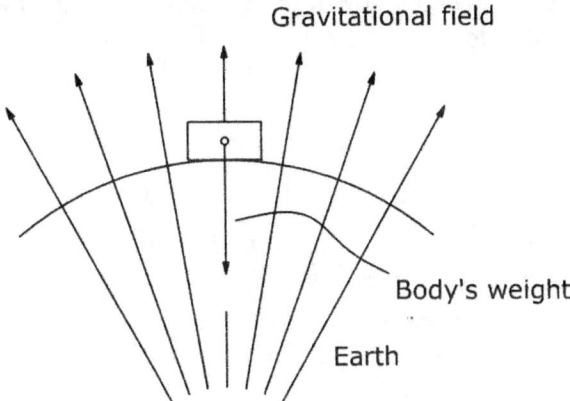

Fig. 2.23. A body at rest or in motion on the surface of the Earth is under inertia or under the gravitational attraction. With the body in motion, it will experience surface friction and air resistance.

A Body Thrown Straight Up from a Standstill

A body (in the presence of gravity) that is thrown straight up from a standstill will fall directly below from where it was originally thrown (Figure 2.24).

This is practically the same as the body at rest. Neglecting the effect of the wind, the ball will be pulled by gravity to where it was originally thrown. (See *A Body Dropped from a Height While Moving.*)

A Body Thrown Straight Up From a Moving Vehicle

A body (in the presence of gravity) that is thrown straight up from a moving vehicle will fall directly below from where it was originally thrown, that is, the vehicle will be away from the ball by the time the ball falls to the ground (Figure 2.25).

The moment the ball is thrown straight up, neglecting the wind, there is only the gravity that is acting on the ball and

THEORY OF GRAVITY

gravity will attract it straight down. (See below *A Body Dropped from a Height While Moving.*)

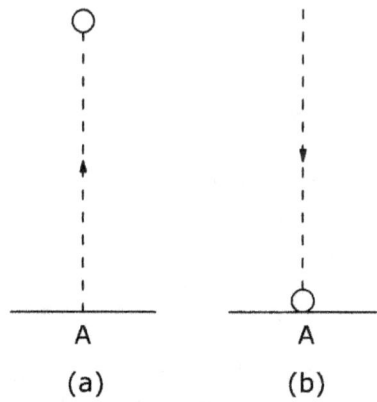

Fig. 2.24. A ball thrown straight from a standstill (a) will fall directly below from where it was originally thrown (b).

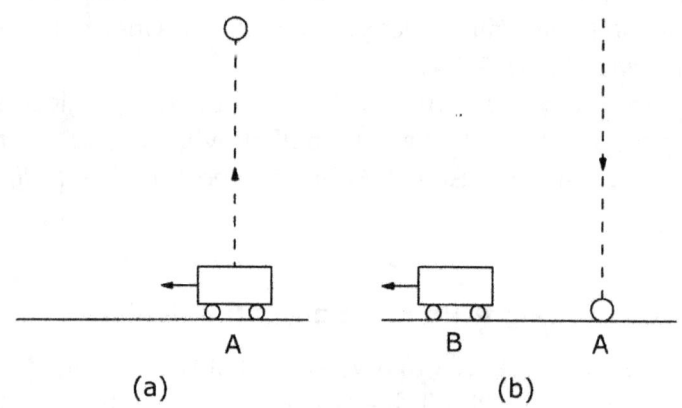

Fig. 2.25. A ball thrown straight up from a moving vehicle (a) will fall directly below from where it was originally thrown (b).

Chapter 2: General Theory of Gravity

A Body Dropped from a Height from a Standstill

A body (in the presence of gravity) that is dropped from a height (neglecting wind) will fall directly below from where it was originally dropped (Figure 2.26).

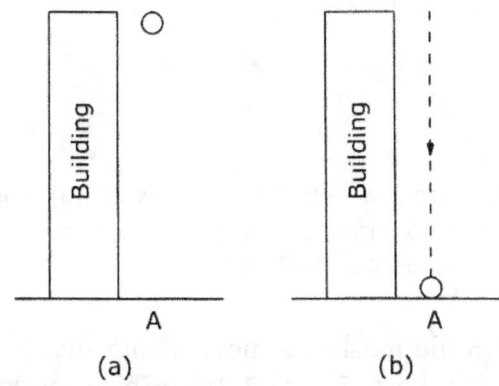

Fig. 2.26. A ball dropped from a height will fall directly below from where it was dropped.

This is the same situation above on *A Body Thrown Straight Up from a Standstill* and *A Body Thrown Straight Up From a Moving Vehicle*. (See below *A Body Dropped from a Height While Moving*.)

A Body Dropped from a Height While Moving

A body (in the presence of gravity) that is dropped from a height of a moving vehicle will fall directly below from where it was originally dropped (Figure 2.27).

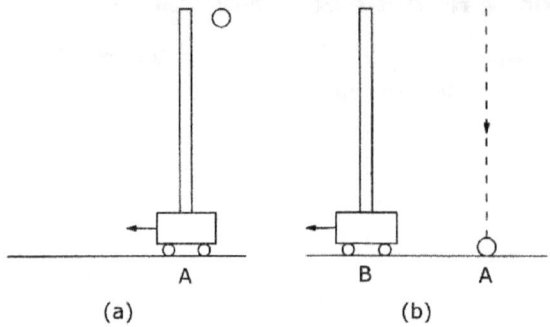

Fig. 2.27. A ball dropped from a height while moving (a) will fall directly below from where it was dropped (b). (Note: Figure 2.27 was drawn for uniformity with Figures 2.24 to 2.26.)

A ball dropped from the mast of a moving ship should land away from the foot of the mast (Figure 2.28). (This experiment is of course affected by the height of the mast and the speed of the ship.) Note that this experiment is in complete agreement with what Galileo had said in his book *Dialogue Concerning the Two Chief World Systems* regarding a stone object dropped from the mast of a moving ship where Galileo said that the ball will fall away from the mast and not at the foot of the mast.[6]

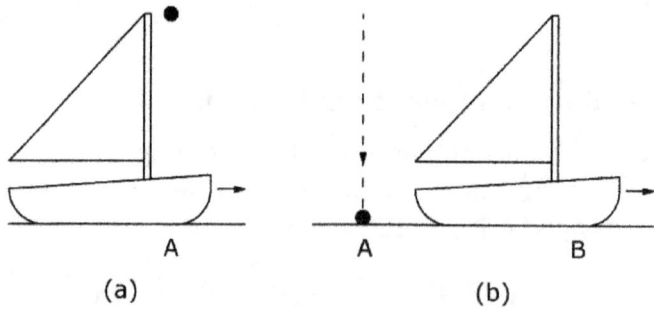

Fig. 2.28. A stone dropped from the mast of a moving ship will land away from the foot of the mast.

Chapter 2: General Theory of Gravity

Note that Figures 2.27 and 2.28 are practically the same demonstration of dropping an object from a moving vehicle.

(This same experiment was corrupted in support of Einstein's theory of special relativity where it says that the observer in a ship will see the ball land on the foot of the ship while an observer on land will see the ball "travel" in a curve to land at the foot of the mast (see Figure 8.1 in Chapter 8). This erroneous adaptation of the experiment to suit Einstein's inertia in his principle of relativity will be used to overthrow his relativity theories in *Chapter 8: Overthrowing Einstein's Relativity Theories*.)

A Body Inside a Ship that is Moving Up Very Fast

A person inside a rocket ship will experience being stuck on his seat (Figure 2.29). This is caused by his weight as he is being pulled down (attracted by gravity) and the direction of the rocket ship going up.

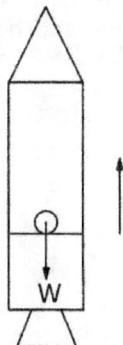

Fig. 2.29. A person in a rocket ship is stuck on his seat as gravity pulls him down while the rocket ship is moving up.

A Body in a Free Fall Inside a Ship

A body in a free fall inside a ship going down is said to experience weightlessness (Figure 2.30). This is not so. A person going down inside a ship that is in a free fall is the same demonstration of Galileo dropping two different weights on a tower. That is, in an ideal situation without air friction (vacuum), the lighter person is going down at the same time as the heavier plane.

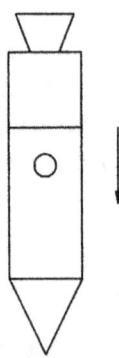

Fig. 2.30. A body inside a ship falling towards the Earth does not experience weightlessness, rather, the body and the ship is falling (in the ideal condition) at the same time. (Hence, there is no arrow showing that the body is going down, as the ship and the body are both going down.)

Chapter 3
Quantum Gravity Theory

Quantum gravity is a field of theoretical physics that tries to unify quantum mechanics and general relativity, that is, it tries to unify the other fundamental forces of physics (strong interaction, weak interaction, and electromagnetic force) with gravity of the theory of general relativity. As a theory involving the fundamental forces where a force has its own mediating particle, it also posits an existence of a theoretical particle called graviton, which mediates the gravitational force. (Some of the theories that try to claim to be quantum gravity theory are string theory and loop quantum gravity. String theory also posits the existence of the graviton.)

Attempts to unify *mathematically* quantum mechanics and general relativity run into a problem of renormalization. (Renormalization is where the sum of all forces does not cancel out and result in an infinite value.) Testing experimentally any theory of quantum gravity runs into a problem since it is said that the energy levels required to observe the conjecture are unattainable in current laboratory experiments.

As we can see from this chapter, physicists had been looking in the wrong places for the quantum gravity theory. (In *Chapter 8: Overthrowing Einstein's Relativity Theories*, we will see why quantum mechanics and general relativity can never be unified.) My quantum gravity theory follows seamlessly my theory of gravity and my theory of everything as relating to the "unification" of all the fundamental forces. The explanation is simple enough, although it needs the understanding and discovery of the nature of the fundamental particles and how they are translated into the creation of gravity.

Micro-scale/Micro-universe (Dipole Magnet Energy Particle): Quarks, Electron, Proton, Neutron, and Atom

A dipole magnet energy particle is an energy particle in which through its spin produces its own energy field. The up quark, down quark, and electron, which are the fundamental particles of matter, are the only dipole magnet energy particle. (Being that the so-called other "generations" of quarks and leptons are just the higher energy particles of the up quark, down quark, and electron.)

Currently, proton is understood to be made up of two up quarks and one down quark while neutron is made up of two down quarks and one up quark. In my quark theory, the up quark is orbited by the down quark. In my New Model of an Atom, proton is orbited by the electron and the nucleus is structured in a proton-neutron pole with the electron orbiting the proton. It is in the structure, mechanism, and configuration of these dipole magnet energy particles from the proton and neutron to the atom that the dipole magnet configuration (see definition below) is maintained.

Definitions for the Dipole Magnet Energy Particle

The following definitions are the scaling of the properties from the dipole magnet energy particles (up quark, down quark, and electron) to the proton and neutron to the atom.

> Dipole magnet energy particle
> > An energy particle, which through its spin produces its own energy field.

Chapter 3: Quantum Gravity Theory

Dipole magnet energy particle structure
  The energy particle and its energy field.

Dipole magnet energy particle mechanism
  The production of the energy field by a dipole magnet energy particle.

Dipole magnet configuration
  The arrangement of one spinning dipole magnet energy particle in the center and the other spinning dipole magnet energy particle being forced into orbit by the central particle.

Dipole magnet structure
  The dipole magnet configuration that resulted overall into the similar dipole magnet energy particle structure.

Up Quark, Down Quark, and Electron

The up quark, down quark, and electron, which are the fundamental particles of matter, are the only dipole magnet energy particle (Figure 3.1).

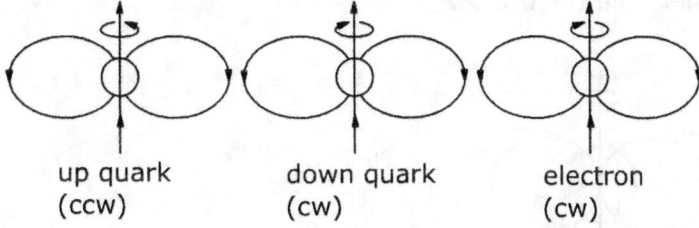

up quark   down quark   electron
(ccw)   (cw)   (cw)

Fig. 3.1. The spinning up quark, down quark, and electron with their energy field.

The dipole magnet energy particle structure is the energy particle with its energy field and its dipole magnet energy particle mechanism is through its spin it creates its own energy field.

(The direction of the spin is the charge of the particle. Looking down, the up quark as a positive charge particle spins counterclockwise while the down quark and the electron as negative charge particles spin in a clockwise direction. The photon of light and the photon of the magnetic field spin in a clockwise direction; hence, they are both negative charge particles.)

Theory of How the Energy Field is Created by a Dipole Magnet Energy Particle

To bring into perspective, the up quark, the down quark, and the electron are considered as a point particle. A point particle is very minute so much so that it does not have any known diameter. So, how does a dipole magnet energy particle produces its own energy field? To answer this question is to "tunnel" into the extremely minute world of the particle and visualize it into our practical world. My theory is that the spinning of the energy particles on its axis throws (emits) out energy at one pole, where this energy that was thrown enters at the other pole (Figure 3.2).

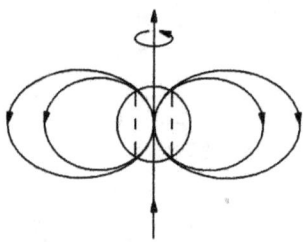

Fig. 3.2. The structure and mechanism of a dipole magnet energy particle that produces its own energy field.

Chapter 3: Quantum Gravity Theory

Imagine spinning a glass with water in it or spinning the water in a glass. The water contained in the glass will naturally rises to the side of the glass through centrifugal force. For the dipole magnet energy particle that is made of energy, spinning on its axis throws out energy particles on its axis forming its energy field that then enters on the other pole.

The direction of spin of these fundamental particles of matter is also their charge. (I had forwarded this in my book *Theory of Everything* my Charge Theory. The same direction of spin also applies to the charge of free particles such as the photon of light, free electron, and free proton.) The up quark spins in the counterclockwise direction, hence we conventionally referred to it as a positive charge particle. The down quark and electron spins in the clockwise direction, hence we conventionally referred to them as negative charge particles. As it is, we can say that whether the energy particle spins in counterclockwise (positive charge) or clockwise (negative charge), the energy field will still come out on top (north pole) and enters at the bottom (south pole). (Note: This assumption of the direction of emission of the energy field is very important in the discussions on charge and parity in *Appendix C: Overthrowing Charge and Parity*.)

One possible way to model the dynamics of a dipole magnet energy particle is to use a clear sphere containing a liquid and spin it on its axis. We can then observe how the liquid behaves inside the sphere. The understanding of the dynamics of the dipole magnet energy particle could also come from the area of plasma physics.

While I had used the term "dipole magnet," the term magnet itself means that it is a dipole magnet. That is, the mechanism of a spinning particle only creates a dipole magnet. A monopole is just not possible as the mechanism of a magnet is through the emission of energy particles or photons from one pole and the going back of the emitted energy particles at the other pole.

THEORY OF GRAVITY

Proton and Neutron

Currently, proton is understood to be composed of two up quarks and one down quark, and neutron is composed of one up quark and two down quarks where all the quarks are lumped together with no structure. In my Quark Theory discussed in my book *Theory of Everything*, the up quark is in the center and is orbited by the down quark (Figure 3.3).

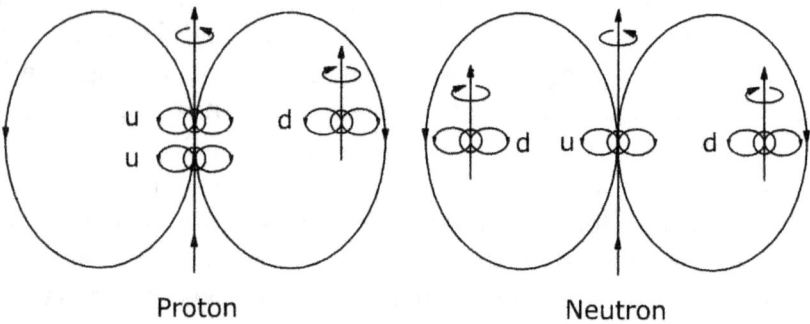

                Proton   Neutron

Fig. 3.3. The structure of the proton and neutron in my Quark Theory. (The spin shown is the direction of the spin of the particle.)

Proton and neutron is understood to be spherical shape. That is, proton and neutron is shape like a particle having a dipole magnet energy particle structure through their energy field. Proton and neutron are in a dipole magnet configuration with the up quark as the center and the down quark as the orbiting particle, preserving the dipole magnet structure.

(Since proton is the only spinning "particle" with its energy field within the atom, it is the only subatomic particle that is able to make the electron orbit it. On the other hand, neutron's up quark and down quarks still spins but its overall energy field created by the quarks does not spin.)

Chapter 3: Quantum Gravity Theory

Atom

An atom is typically made up of proton, neutron, and electron. Hydrogen atom has only one proton and one electron, while the rest of the elements have the proton, neutron, and electron. (When neutron is paired with the paired proton and electron, then the extra numbers of neutron will be the different isotopes of the element.) In my theory of a New Model of an Atom discussed in my book *Theory of Everything*, proton and neutron are arranged in a pole and that proton is orbited by only one electron (Figure 3.4).

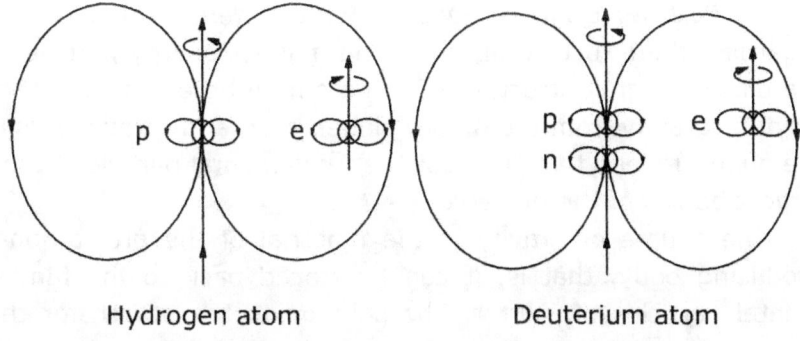

Hydrogen atom　　　　　Deuterium atom

Fig. 3.4. The structure of the atom represented by the hydrogen atom and its isotope deuterium showing the position of the neutron.

With the proton at the center of the atom orbited by an electron and the nucleus is a pole of proton and neutron stacked on top of each other, the atom is in a dipole magnet configuration preserving the dipole magnet structure. The atom is typically viewed as spherical shape, although it should be noted that with the stacked proton and neutron nucleus and the configuration of the electrons, the atom might not be completely spherical in shape.

Macro-universe (Quantum Leap: Dipole Gravitational Body): Planet, Star, and Galaxy

How do we "unify" quantum mechanics with gravity that leads to quantum gravity theory? Remember that the purpose of quantum gravity theory is to explain the world of the atoms and the universe or explain everything from the atoms to the universe. The solution to a "quantum gravity theory" does not deal with the current approach of the use of mathematics or in simulating the Big Bang but rather with the discovery of the nature of the fundamental particles of matter discussed in the above section on *Microscale/Micro-universe (Dipole Magnet Energy Particle: Quarks, Electron, Proton, Neutron, and Atom)* regarding the structure of the dipole magnet energy particle to the dipole magnet structure of the atom and the structure that produce gravity from the moon, planet, star, and galaxy. This is the quantum leap that "unifies" the fundamental particles to the largest bodies of the universe

The source of gravity is the material of the gravitational producing body, that is, it can be traced back to the fundamental particles of matter. The solution to the search for the quantum gravity theory is the understanding of the creation of energy field (magnetic field) of the dipole magnetic energy particle and the energy field (gravitational field) of the gravitational producing body.

Definitions for the Dipole Gravitational Producing Bodies

The following definitions are the scaling of the properties (structure and mechanism) of the gravitational producing bodies from the moon, planet, star, planetary system, solar system, and galaxy.

>Gravitational Producing Body.
>> A body such as the moon, planet, and star that produces their gravitational field. (Note that just as a

solar system has a star at its center, a galaxy should also have a massive central star or at least the mechanism that produces an energy field.)

Dipole gravitational body structure
　　The structure of a gravitational producing body that produces the gravitational field.

Dipole gravitational body mechanism
　　The production of the gravitational field by a gravitational producing body.

Dipole gravitational configuration
　　The arrangement of one spinning body in the center and the other spinning body being pushed into orbit by the central body such as the planet and its moon, the star and its planet, and the galaxy's central star and its material. (Note that this is practically similar to the *Dipole Magnet Configuration*.)

Dipole gravitational structure
　　The dipole gravitational configuration that result overall into the same as the dipole gravitational body structure. (Note that is practically similar to the *Dipole Magnet Structure*.)

Moon, Planet, and Star

The commonality regardless of the material composition of the moon, the planet, and the star is that they all spin and produces their own gravitational field (Figure 3.5)

　　That is, the structure of these gravitational producing bodies produces gravity. Through their dipole gravitational body structure, the dipole gravitational body mechanism produces gravity.

THEORY OF GRAVITY

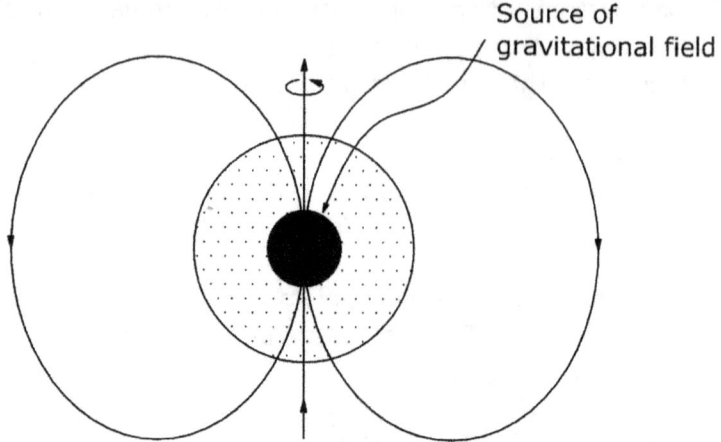

Fig. 3.5. The moon, the planet, and the star spin and produces gravitational field.

Theory of How Gravity is Created by the Star, Planet, or Moon

How is gravity created? The answer often cited is Einstein's general theory of relativity, which states that gravity is the warping of fabric of spacetime by a massive body. That is, gravity is not a force at all, which is at odds with how we experience gravity as a force.

So, how is gravity produced by a gravitational producing body such as the moon, the planet, and the star?

The core of the Sun is about 0.20 to 0.25 of its radius. It is the hottest part of the Sun and is made of hot dense gas in plasma state. The core produces almost all of the Sun's heat by nuclear fusion where hydrogen is converted to helium.

Our solar system is currently known to have nine planets. Mercury, Venus, Earth and Mars are solid planets and are all considered to have a solid core. Earth is known to have a solid inner core made of iron and nickel and a liquid outer core. It is said that the convection of the liquid metals in the outer core

Chapter 3: Quantum Gravity Theory

creates the Earth's magnetic field.[1] Jupiter is considered to have a rocky core and parts of metallic hydrogen, liquid hydrogen, and helium gases. Saturn is considered to have a rocky core and parts of metallic hydrogen, hydrogen and helium gases, and liquid hydrogen and helium. Uranus is considered to have a rocky core; a mantle of water, methane, and ammonia ices; and an atmosphere of hydrogen, helium, and methane. Neptune is considered to have a rocky core; a mantle of water, methane, and ammonia ice; and an atmosphere of hydrogen, helium, and methane. Jupiter, Saturn, Uranus, and Neptune are known to have rings, although Saturn is more famous for its rings. Pluto is considered to have a rocky core, a mantle of ice, and methane frost. The Moon is considered to have a core, a mantle, and a crust. It is believed to have a solid inner core and a liquid outer core.

As I have reached this point in my discussion of gravity, we should already know that gravity or gravitational field and magnetic field are practically the same as their mediating particle is photon.

So how is gravity created? In the Sun, there is a portion of the core made of plasma. In the Earth there is the solid inner core and hot liquid metallic outer core. In other planets, it is said to be made of rocky core, although I suspect they should have also a hot liquid outer core. Our moon is also believed to have a solid inner core and liquid outer core. (It was observed that the Moon is losing its hot liquid core to the solid core, which could be the explanation for its gravity getting weaker.) We can observe that there is a commonality among the stars, planets, and moons: they all have a solid inner core of some kind and a usual hot liquid outer core for the planets and the moon, and they all spin. There are three possible theories on the creation of gravity.

For the moon and the planet, one way the gravity is created is by the rubbing of the hot liquid outer core to the solid inner core magnetizing the inner core (Figure 3.6).

THEORY OF GRAVITY

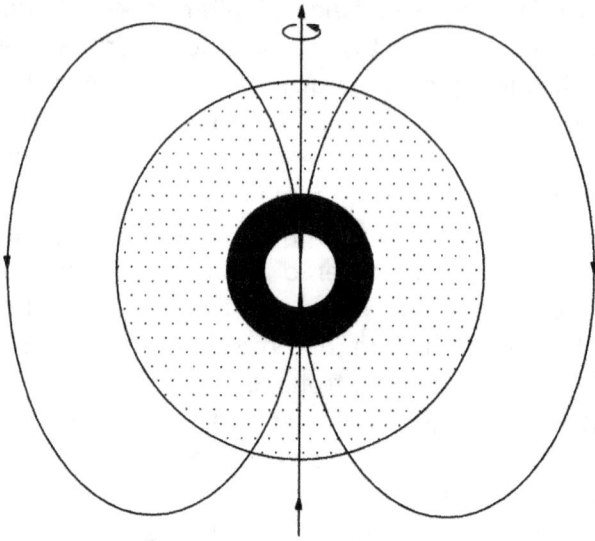

Fig. 3.6. The hot liquid outer core rubs with the solid inner core magnetizing the solid inner core where the polarity created is along the spin axis.

The solid inner core acquires a magnetic property just like a metal acquires a magnetic property when rubbed with another metal. That is, the atoms of the metal could be aligned by rubbing it with another metal so that the whole metal becomes a magnet. That is, gravity here is that of magnetic field coming from the atom.

For the moon and the planet, gravity can also be created by electromagnetism. We know that when a magnet is moved in and out of a wire conductor coil that a current is created on the wire conductor. The way the magnet produces electrical current on the wire conductor is that the magnet's magnetic field acts like a sweeper on the electrons of the conductor forcing the electrons to flow. To reorient another way, the moving hot outer core provides the photons (as heat) that sweeps the electrons of the solid inner core creating an electric current in a part of the solid inner core like a toroid (Figure 3.7).

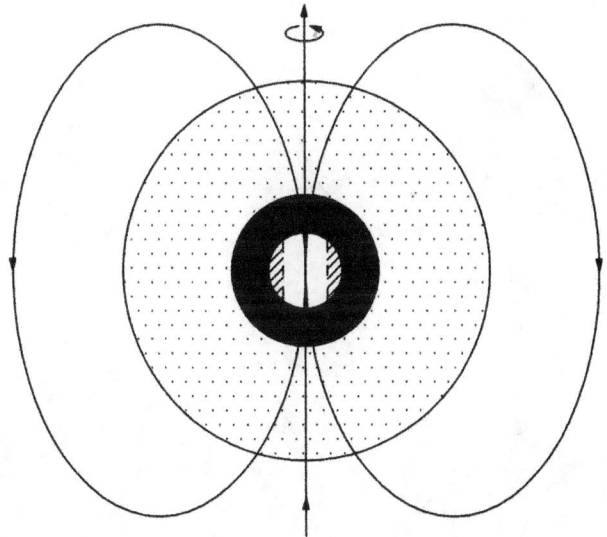

Fig. 3.7. The hot liquid outer core of the moon and the planet produces photons (heat) that act as a sweeper (much like the magnetic field of a magnet) to the electrons of the solid inner core. The part of the solid inner core affected by the electric current is shape like a toroid, forming the body's gravitational (electromagnetic) field.

For the moon, the planet, and the star, gravity can also be created by their spin and the heat imparted on the atoms on their solid inner core (Figure 3.8). What the spin does is to align the poles of the atom and the what the heat does is to impart these excess photons on the atom's magnetic field and just like the creation of the energy (magnetic) field of the dipole magnet energy particle and that the excess energy is emitted on one pole where the whole atoms of the core acts as one just like a magnet to create the body's gravitational field.

THEORY OF GRAVITY

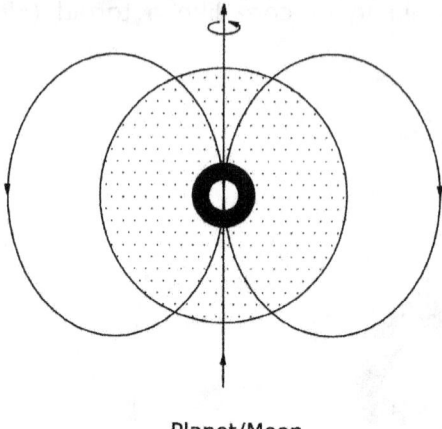

Planet/Moon

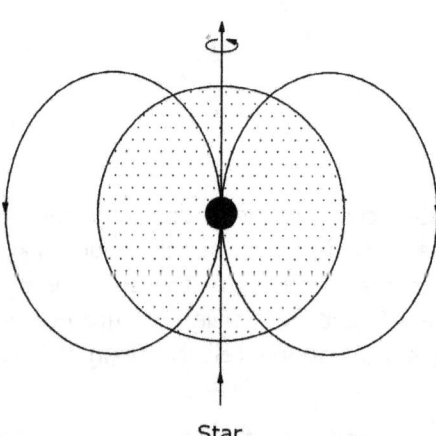

Star

Fig. 3.8. The gravitational producing body's (moon, planet, and star) spin aligns the atoms of the inner core and the heat imparted on the atom makes the atom emit stronger energy (magnetic) field where the whole core acts as a giant magnet producing the body's gravitational field.

Note that experiments at a practical scale can be done to verify the aforementioned theories.

Chapter 3: Quantum Gravity Theory

Theory of How the Spin of the Moon, Planet, Star, and Galaxy is Started

My theory of how the spin is started by the moon, the planet, and the star is that the impetus for the spin and direction of their spin has its source in the spin and direction of spin of the up quark (counterclockwise spin/positive charge) that was passed down to the proton and to the atom. (The spin of the galaxy is derived from the spin of its central star.) When the moon and the planet formed with their solid inner core and hot liquid outer core, it seems that from the atom in the center of the core the spin of the up quark is translated outwards to the whole body to spin it in the direction of the spin of the up quark. This is the answer to the questions of Galileo, Descartes, and Newton on how or why the planets orbit around the Sun and the moon orbit around a planet.

Solar System and Planetary System

The Sun (star), which is the center of our solar system, forces the planets to orbit around it. The Sun as the center with the planets orbiting around it is configured like a dipole magnet (Figure 3.9).

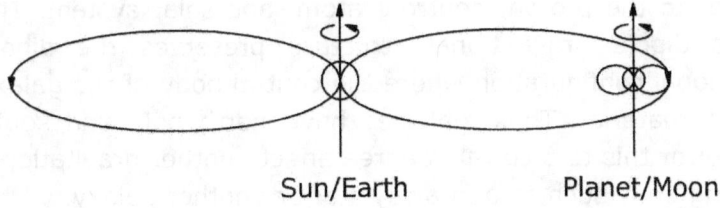

Sun/Earth   Planet/Moon

Fig. 3.9. The solar system and planetary system are configured like a dipole magnet. (The spin shown is the direction of the spin of the bodies.)

The dipole gravitational structure preserves the dipole gravitational configuration where a spinning central body makes another spinning body orbit around it. It is the immense gravitational field of the central body that dominates the whole system to produce a dipole gravitational structure similar to that of the proton, neutron, and the atom.

We can also observe that there is also the planetary system where the moon orbit around a planet. For the giant gaseous planets in our solar system (Jupiter, Saturn, Uranus, and Neptune), we will notice that aside from their moons that orbit them, they also have the rings which corresponds to the asteroid belt of our solar system. That is, just like the Sun that is primarily made of the hydrogen atoms, it could be a property of a gaseous gravitational producing body. Otherwise, it could be the property of a sizeable gravitational producing body (of which there is a great chance is primarily gaseous).

Galaxy

A nebula is typically made of gas, dust, and the products of supernovae. When a massive star is formed that dominates all the bodies in the nebula, the birth of a galaxy starts. The galaxy typically starts as a spherical galaxy (which means that it is already spinning as a whole) until it becomes an elliptical galaxy (Figure 3.10). Having a massive central star, the galaxy is similar to the proton, neutron, atom, and solar system. The galaxy's dipole gravitational structure preserves the dipole gravitational configuration where the central body of the galaxy spins its material. Thus, galaxies have north pole and south pole and for this to a certain degree affect another gravitational producing body such as a faraway star or another galaxy.

(The theory of general relativity posits a black hole that is caused by a massive body. Currently, it is being perpetuated in the science publications that a black hole is a hole or that there is a black hole in the center of a galaxy. If it were, our solar system, the spiral galaxies, and elliptical galaxies should have the structure of a funnel. Einstein's special relativity and general relativity theories will be overthrown in Chapter 8.)

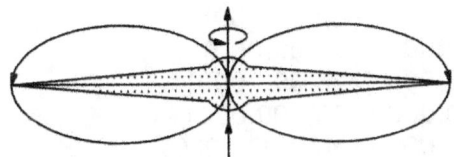

Fig. 3.10. The galaxy and its dipole gravitational structure. (The spin shown is the direction of the spin of the galaxy.)

Bode's Law: The Distance of the Planets from the Sun

Our solar system is composed of the following bodies (starting from the Sun): Mercury, Venus, Earth, Mars, asteroid belt, Jupiter, Saturn, Uranus (discovered in 1781), Neptune (1846), Pluto (1940), and Kuiper belt (1992). Johannes Kepler (1571-1630) suspected that there must be a planet between Mars and Jupiter, which is occupied by the asteroid belt.

In 1772, Johann Elert Bode (1747-1826), the director of Berlin Observatory, called into attention a strange "law" concerning the distances of the planets that would be called Bode's law. (This "law" was discovered some years earlier by another German, Johann Daniel Titius of Wittenberg.)

A sequence of number is as follows: 0, 3, 6, 12, 24, 48, 96, and 192. Notice that apart from the first number, the next number is multiplied by 2 to get to the next number sequence, and so on. Now, add 4 to the number to the number sequence to derive a new sequence: 4, 7, 10, 16, 28, 52, 100, and 196. Assign this number sequence to the planets: 4(Mercury), 7(Venus), 10(Earth), 16(Mars), 28(-), 52(Jupiter), 100(Saturn), and 196(Uranus). Let us look at the actual distance of the planets from the Sun (Table 3.1).

THEORY OF GRAVITY

Table 3.1 The distance of the planets from the Sun.

Planets	Distance in miles (in kilometers)
Mercury	35,900,000 (57,800,000)
Venus	67,200,000 (108,000,000)
Earth	93,000,000 (150,000,000)
Mars	142,000,000 (228,000,000)
Jupiter	483,000,000 (778,000,000)
Saturn	885,500,000 (1,400,000,000)
Uranus	1,800,000,000 (2,900,000,000)

To find the planet's distance from the Sun: Take the number of the planet according to Bode's law, say, Earth which has a value of 10, divide this number by 10, and multiply the result by 93,000,000. This gives us 93,000,000 miles, which corresponds to the actual distance the Earth to the Sun. Now take Venus, which is 7, divide it by 10, and then multiply by 93,000,000. This gives us 65,000,000 miles, which is quite close. Table 3.2 shows the closeness of the values of Bode's law and the actual value of the computation of the planets distances (Actual value = (distance of the planet x 10)/93,000,000).

Table 3.2. Bodes law on the distances of the planet from the Sun

Planet	According to Bodes law	Actual value
Mercury	4	3.9
Venus	7	7.2
Earth	10	10
Mars	16	15.3
-	28	-
Jupiter	52	52
Saturn	100	92
Uranus	196	193.5

Chapter 3: Quantum Gravity Theory

Notice that there is a missing "body" between Mars and Jupiter, which resulted in the hunt for a "planet" that has a Bode's law number of 28. This area of the supposed "planet" happened to be occupied by the asteroid belt. The result was the finding of the large bodies called Ceres and Vesta.

Why are the Planets Located at their Distance from the Sun: Periodic Table of the Elements and the Gravitational Field

Bode's law showed us that there is a very uncanny mathematical pattern in the distance of the planets and other bodies (asteroid belt and Kuiper belt) from the Sun. The answer to this mathematical order could only be caused by the aspect of the Sun's gravitational field.

The gravitational field of the Sun arranges the bodies of the solar system to their distances from the Sun and their physical and (to possibly some extent) chemical structure (solid, unformed asteroid belt, gaseous): Mercury, Venus, Earth, and Mars are solid planets; unformed materials composing the asteroid belt; Jupiter, Saturn, Uranus, and Neptune are gaseous giant planets; and then the planet Pluto in the Kuiper belt (Figure 3.11).

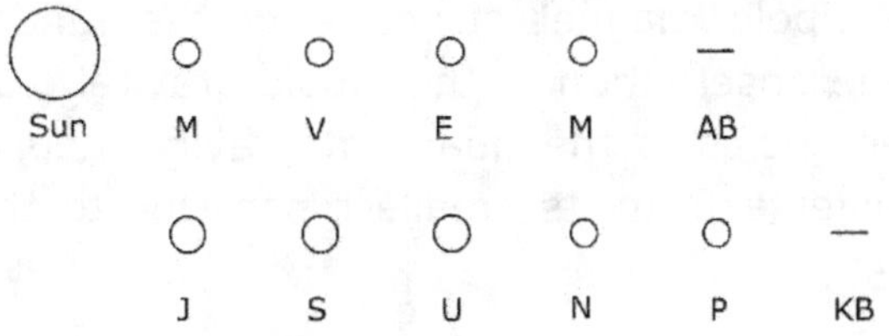

```
 ◯    o    o    o    o    —
Sun   M    V    E    M    AB

 O    O    O    O    O    —
 J    S    U    N    P    KB
```

Fig. 3.11. The pattern of our solar system. (AB=Asteroid belt, KB=Kuiper belt)

We can also observe the same formation of the moons as the planets, and asteroid belt as the rings in Jupiter, Saturn,

Uranus, and Neptune. That is, what is called belt in our solar system is called rings in the gaseous planets.

In my book *Theory of Everything*, I forwarded the structure of an atom in my New Model of an Atom: One electron orbiting one proton and that the proton and neutron are arranged in a pole. What the New Model of an Atom showed was that the magnetic field of the proton arranges the electron in what is called an electron configuration to give an element the specific properties (Figure 3.12).

What can be observed in the periodic table of the elements is that the elements are grouped according to metals, gaseous, nonmetals, and semimetals (Figure 3.13). (Periodic after all means that something that is repeating.)

In the periodic table of the elements the only simple answer to the repeated groupings of the elements with certain characteristics can only be the proton's (as a nucleus) magnetic field. While the solar system and planetary system is only similar in structure to an atom with its nucleus, the only possible answer to the structure of the solar system or the planetary system could only be the Sun's and planets' gravitational field, respectively.

Quantum Gravity Theory in a Nutshell

The discovery of the dipole magnet structure of the fundamental particles and the observation of the dipole gravitational structure of the larger bodies is the quantum gravity theory, which explains the universe from its smallest particles to the largest bodies.

Chapter 3: Quantum Gravity Theory

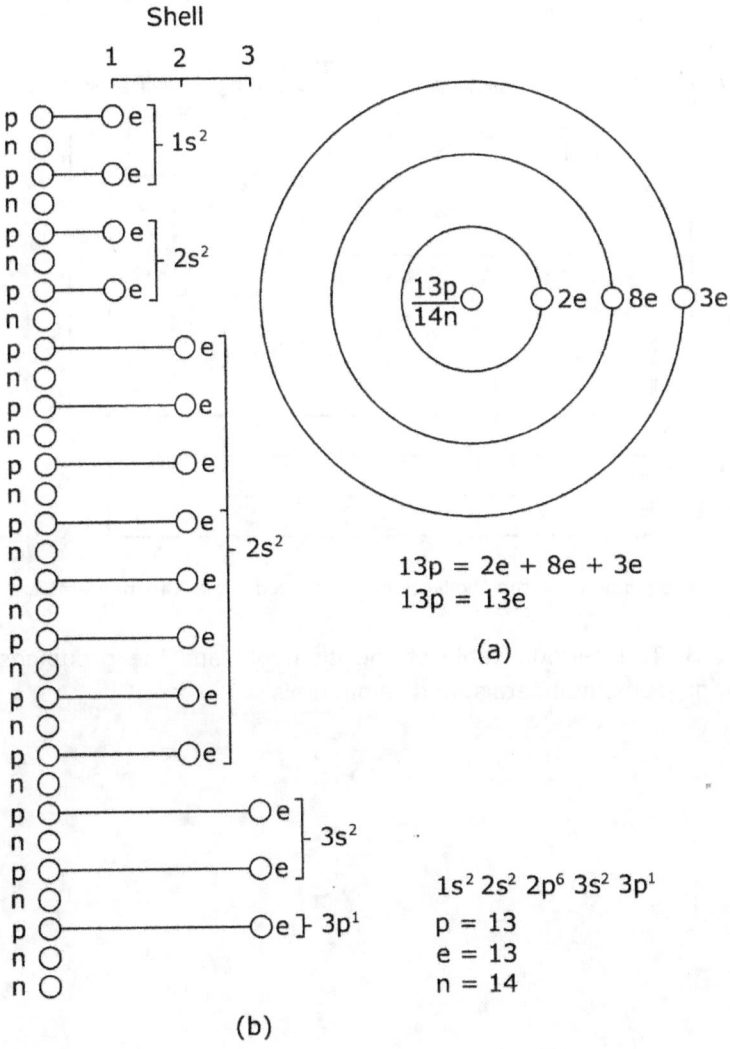

Fig. 3.12. Each addition of the proton and electron arranges the electrons of the element to give specific properties to that element. a) Top view. The distribution of the electrons in an aluminum atom. b) Side view. A simple Christmas tree structure of an aluminum atom.

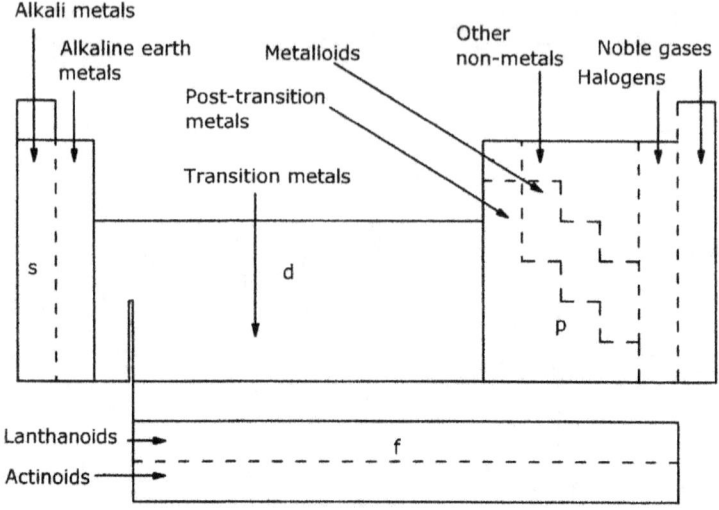

Note: Hydrogen atom on the Alkali column is grouped in the "Other non-metals"

Fig. 3.13. The periodic table of the elements and the groupings of metals, gaseous, nonmetals, and semimetals.

Chapter 4
Theory of Everything: Standard Model with Gravity

This chapter is the fulfilment of my theory of everything I had forwarded in my book with the same title. My theory of everything includes gravity in the standard model and the "unification" of all the fundamental forces. (It is said that the standard model is the theory of *almost* everything except that it cannot explain gravity.) Among the many subjects in physics that my theory of everything can explain are the following:

- It is based on the standard model and corrected quantum mechanics. (The correction to the quantum mechanics includes the removal of the wave property of the particle as a particle has only a particle property and quantum entanglement as the interactions of the two particles is limited only by the reach of their energy field.)
- It can explain the source of mass of the particles. (Mass is an inherent property of a particle in motion ($m=E/c^2$) and is not caused by the Higgs mechanism.)
- It can explain gravity based on the standard model.
- It can explain gravity without the theoretical mediating particle called graviton. (The existence of graviton is also important to the string theory.)
- It can explain dark matter and dark energy.

My theory of everything explains in simple terms our world based on the fundamental particles (nature's basic building blocks) and the fundamental forces. This chapter is the culmination of what I had discussed in my book the *Theory of Everything* with the simplification of the standard model.

* * *

Note that the subject of the energy field will be discussed repeatedly. It will be discussed first under the section *Fundamental Particles* on subsection *Energy Field* to show that all the fundamental forces came from the fundamental particles (the dipole magnet energy particles: up quark, down quark, and electron) and that nothing exist without them, that is, no energy field exist without them. (The energy field here refers to the currently recognized fundamental forces of strong interaction, magnetism, electromagnetism, and gravity. Note that the Higgs field is not even considered as one of the fundamental forces and rightfully so because it does not exist. The Higgs boson will be discussed in *Chapter 9: Overthrowing Higgs Boson, String Theory, and the Question on the Dominance of Matter over Antimatter in the Universe.*) Under the section *Fundamental Forces* on subsection *Energy Field: Preliminary* will be discussed the confusion on what is strong interaction, magnetism, electromagnetism, gravity, and the nature of light as an electromagnetic wave. Under section *Summary: Energy Fields, Emissions, and the Unification of the Fundamental Forces* will be discussed the commonality of the fundamental forces of strong interaction, magnetism, electromagnetism, gravity, and light; and the commonality of light and radioactivity that will led to the understanding of the "unification" of the fundamental forces in the standard model.

Chapter 4: Theory of Everything: Standard Model with Gravity

Fundamental Particles

The fundamental particles of the standard model (up quark, down quark, and electron) are the building blocks of our physical world. A fundamental particle is described by its structure, mechanism, and properties. Its structure is that it is an energy particle and has an energy field. Its mechanism is that it is spinning and that through its spin it creates its energy field. Its properties are that its direction of spin is its charge and its motion gives it its mass.

Note that unlike the current standard model that includes neutrino as a fundamental particle, the neutrino in my standard model is under the fundamental force of radioactivity (formerly weak interaction). Neutrino is an emission resulting from the neutron changing into a proton and vice versa. Both the photon of light and the neutrino of radioactivity are an emission of the atom. The photon of light is a (negative) charge particle since it spins, while the neutrino has no charge since it does not spin.

Dipole Magnet Energy Particle

A fundamental particle is a dipole magnet energy particle, which by its motion of spinning in place as a center (up quark) or spinning and orbiting around a center (down quark and electron), creates its own energy field. (Its shape must be spherical due to its spinning motion just as large bodies (such as the star, planet, and moon) that are spinning are mostly spherical.) The up quark, down quark, and electron, which are the fundamental particles of matter, are the only dipole magnet energy particles (Figure 4.1).

Charge/Spin (Charge Theory)

As stated in my Charge Theory in my book *Theory of Everything*, spin and the direction of spin is the charge of the particle. That is, spin and charge are both the same. Positive charge particle spins in a counterclockwise (ccw) direction, negative charge particle spins in a clockwise (cw) direction, and neutral charge particle does not spin. (Refer to Figure 4.1 for the direction of the spin as the charge of the charge particles. Neutrino particle does not spin. Neutron's up quark and down quark spins but its resultant energy field does not spin.)

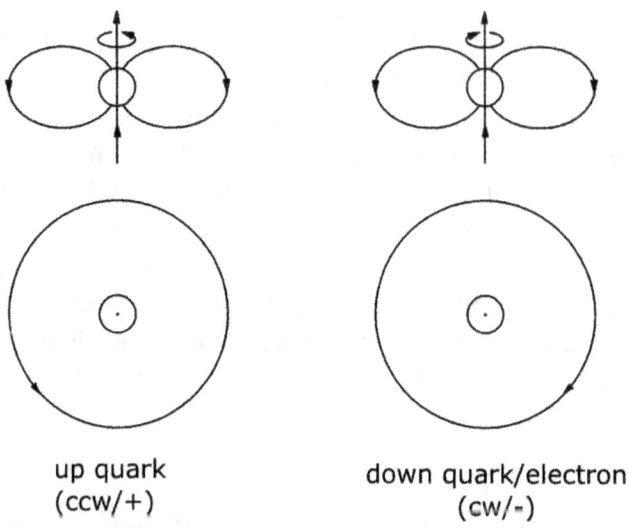

up quark　　　　　　　down quark/electron
(ccw/+)　　　　　　　　(cw/-)

Fig. 4.1. Dipole magnet energy particles.

Energy Field

The spinning motion of a dipole magnet energy particle produces its energy field (refer to Figure 4.1). Bound particles (particles within the atom: up quark, down quark, proton, neutron, and electron) emits an energy field. Their energy field can be called magnetic field since it is their resultant energy field that created the magnetic field of the magnet—proof that the atom has an energy field all the way to the fundamental particles. (We can also say that it is the energy field of the dominant up quark inherited by the proton and the atom that produces the magnetic field of the magnet.) The only other two

Chapter 4: Theory of Everything: Standard Model with Gravity

energy fields are from the free electrons (electricity) moving in a conductor called electromagnetic field and the gravitational field.

Mass (Mass Theory)

Mass is an inherent property of a particle derived from its motion ($m=E/c^2$), whether by spinning in place as a center (up quark) or spinning and orbiting a center (down quark and electron). Proton and neutron derived their resultant mass from their quarks.

Whereas Einstein's equation $E=mc^2$ is explained as the equivalence of matter and energy, I had shown that it is the missing understanding of the source of mass in the standard model. In $E=mc^2$, E is the energy of the particle, c is the motion of the particle, and m is the mass. That is, the mass of the particle depends on its energy and motion (speed). (Technically, we cannot convert mass or matter into energy or vice versa since everything is energy. Rather, what is thought of that was converted from "mass" to "energy" is the energy released by the changing of the neutron to proton.)

For a free particle that is moving in space (photon, neutrino, free electron, and free proton), it derives its mass from its momentum. In the formula $p=mv$: where p is the momentum, m is the mass, and v is the velocity.

Fundamental Forces

The current fundamental forces are the following: strong interaction, weak interaction, electromagnetism, and gravity. Currently, the theory of the standard model involves only the strong interaction, weak interaction, and electromagnetism as it does not include gravity.

Presently, physicists are trying to "unify" all these forces through mathematical means. The "mathematical means" here was based on the idea that a short period during the Big Bang, the fundamental forces were thought to be unified like a soup of energy that would later separate from each other.

The first often cited unification was that of electricity and magnetism, which was said to have been unified by James Clerk Maxwell in his electromagnetic theory. Electromagnetism and weak interaction was also said to have been unified into electroweak theory. In my book the *Theory of Everything*, I said that there is actually no unification of the fundamental forces but rather *it is the understanding of what and where those forces are and how they operate*. Thus, any purported unification of the fundamental forces especially through mathematical means is not only highly questionable but also outright wrong as we shall see from the discussions below. This chapter is the completion of what I had started in my book *Theory of Everything*, which includes gravity. How the gravity operates was discussed in *Chapter 2: General Theory of Gravity* and *Chapter 3: Quantum Gravity Theory*.

In my theory of everything, I had reorganized the fundamental forces by categorizing them into two: the fundamental forces of matter and the fundamental forces in nature. The fundamental forces of matter are within the atom: quark strong force, quark weak force, atomic strong force, and atomic weak force. (The quark strong force or the resultant quark strong force and the quark weak force is the strong interaction. The atomic strong force or the combination of the atomic strong force and the atomic weak force is the strong nuclear force. This will be made clear down below.) The fundamental forces in nature are electromagnetism (electricity), magnetism (atom), light, gravity, and radio-activity.

It will be noticed that quark strong force, quark weak force, atomic strong force, atomic weak force, electromagnetism, and magnetism involves a particle (up quark, down quark, electron, proton, neutron, and the atom) and the particle's energy field. Gravity is related to the aforementioned forces in that gravity came from the fundamental particles. Light's photon and

Chapter 4: Theory of Everything: Standard Model with Gravity

radioactivity's neutrino are related, as they are emissions from the atom.

Strong Interaction

There is often a confusion with the strong interaction and the strong nuclear force. Currently, strong interaction is understood to be the mechanism responsible for the strong nuclear force (also called strong force or nuclear force). It is said that on a larger scale, strong interaction binds the protons and neutrons in an atom, while on the smaller scale strong interaction binds the quarks inside the proton and neutron.

Strong nuclear force is also said as a residual effect of the strong interaction. For this matter, strong nuclear force is also called residual strong force. (The proper term for strong nuclear force as it relates to strong interaction should be residual strong interaction.) The subject of strong interaction has come to mean quantum chromodynamics; hence, the term color force is used to describe the interaction between the quarks. In the strong interaction, the mediating particles for the quarks are the (colored) gluons while in the strong nuclear force the mediating particle is the pion. (Pion is a meson, which is made up of two quarks: a quark and an antiquark. There are three known pion: neutral pion, positive pion, and negative pion.)

In the following sections on *Quark Strong and Weak Force* and *Atomic Strong and Weak Force* in the clarification of strong interaction and strong nuclear force, it will be understood why the strong nuclear force is also called the "residual effect" of strong interaction.

Energy Field: Preliminary

In the current fundamental forces it is not completely known where is the strong interaction within the quarks, where is the strong nuclear force (also called as residual strong force) within the atom, or what is the connection between the strong interaction and strong nuclear force.

There is the confusion on what is the strong interaction and magnetism as magnetism (from the magnet) also came from the atom. There is a confusion with electromagnetism and magnetism as both interact with each other and that even most of the time they are both used interchangeably. (It is a fact that magnetism is not considered as a fundamental force in the standard model as it is lumped in electromagnetism.) There is the confusion with gravity and magnetism. Magnetism attracts and repels while gravity is thought only to attract (this was discussed in *Chapter 2: General Theory of Gravity*). Earth's gravitational field is often called magnetic field as it is observed to affect the magnetic compass. (It is never thought that gravity repels even when that is how the magnetic compass works—the North Pole of the Earth repels the north pole (labelled S) of the magnetic compass for the compass to attract the south pole (labelled N) for the magnetic compass to "point" to the North Pole.) There is also confusion with the nature of light and electromagnetic waves.

The confusion basically stem from not understanding what and where these fundamental forces are and how they operate. It is in the proper identification of the energy field of the fundamental forces that leads to the "unified" understanding of the fundamental forces. The correct understanding of the fundamental forces simplifies our understanding of nature that leads to the theory of everything and the simplification of the standard model.

Quark Strong and Weak Force (Quark Theory): Strong Interaction

The current understanding of the quarks within the proton and neutron is that they have no structure. In my Charge Theory, I had forwarded that there is a structure of quarks within the proton and neutron to explain why the proton is a (positive)

Chapter 4: Theory of Everything: Standard Model with Gravity

charge particle (spinning in a counterclockwise direction) and why the neutron is a particle with no charge (not spinning). The understanding of the strong interaction is the understanding of the structure of the energy field of the quarks. The quark strong force corresponds to the polar magnetic field while the quark weak force corresponds to the field spin magnetic field (Figure 4.2).

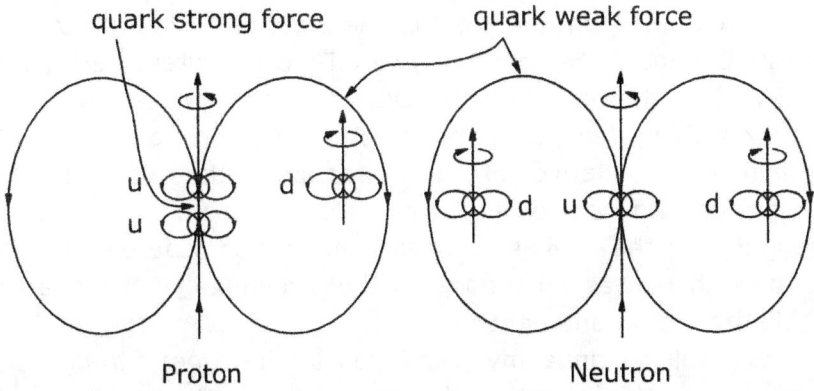

Fig. 4.2. Structure of quarks within the proton and neutron showing where the quark strong force and the quark weak force.

The quark strong force can be seen in the structure of the proton as the polar magnetic fields of the two up quarks that are holding them both together one on top of the other. The quark weak force is referred to as the force that is exerted by the up quark to hold and make the down quark orbit around it. The quark weak force is by no means a "weak" force in the sense that the quarks within the proton require more energy to break it apart (compared to the subatomic particles of the atom).

It can be argued also that it is only the quark strong force (the polar magnetic field) that can be called strong interaction even though the forces that govern within the quarks of the proton and neutron are consists of the quark strong force and quark weak force (but then again, the quark strong force and the quark weak force are the same structure of the energy field of the up quark), as it is the quark strong force that will be passed down to the energy field of the proton.

In the standard model, the mediating particle (gauge boson) for the strong nuclear force is the gluon. My discovery of the structure of the quarks inside the proton and neutron shows that there are no gluons since the mediating particle of the energy field from the atom to the proton and neutron to the quarks is photon. Hence, my Quark Theory refutes the theory of quantum chromodynamics (QCD). When I wrote my book *Theory of Everything*, I had been bothered by how odd the QCD explained the dynamics of the interaction of the quarks inside the proton and neutron in terms of color schemes. That is, I thought that the use of color was an ad hoc scheme created because there was no theory on the structures of the quarks inside the proton and neutron.

As I will continue my discussion in the *Atom Strong and Weak Force (New Model of an Atom): Strong Nuclear Force or Residual Strong Interaction*, my Quark Theory and New Model of an Atom will explain why our current strong nuclear force is referred to as the "residual effect" of strong interaction.

Atom Strong and Weak Force (New Model of an Atom): Strong Nuclear Force or Residual Strong Interaction

The current understanding of the subatomic particles inside the atom is that the protons and neutrons are lump as a nucleus and that the electrons orbit the nucleus in a random order. In my New Model of an Atom I had forwarded in my book *Theory of Everything* the proton and neutron are arranged in a pole and (that in a stable atom) only one electron orbits a proton. In this structure of the atom based from the New Model of an Atom, the atom strong force (polar magnetic field) holds the nucleus together while the atom weak force (field spin magnetic field) of the proton holds the electron (Figure 4.3).

Chapter 4: Theory of Everything: Standard Model with Gravity

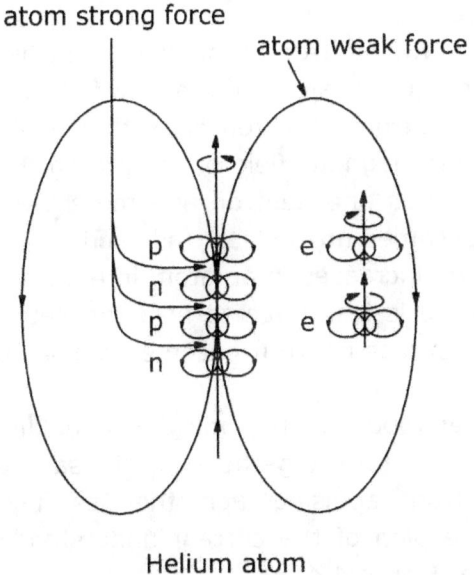

Helium atom

Fig. 4.3. Atom strong force and atom weak force in the helium atom.

In this New Model of an Atom, we can see that it is the same force in the quark strong force that is holding the protons and neutrons together in a pole (the resulting polar magnetic field of the proton and neutron) and the resultant force (the resulting field spin magnetic field) of the quark weak force of the proton that is holding the electron and making it orbit around the proton. We can see from Figure 4.3 that the polar magnetic field of the up quark (or the resultant magnetic field of the up quark and down quark) that was inherited by the proton is now proton's polar magnetic field that is the one holding the nucleus together. The resultant magnetic field of the protons and electrons (much like the up quark and down quark that made up the proton and neutron) would become the magnetic field of the atom. This is why the strong nuclear force is also called residual strong interaction, as it is the polar magnetic field of the up quark that is handed down as the polar magnetic field of the proton.

Physicists wondered why neutron becomes stable when it is inside the stable atom (especially an atom with an equal number of proton and neutron). The reason is that neutron is held in place by the polar magnetic field of the proton. For the most part, one proton holds one neutron and more than that and *usually* the atom becomes unstable or radioactive. That is, as the number of neutron increases in an atom in excess of the proton-neutron pair, there is less force to hold the neutron in the nucleus and so the excess neutron becomes radioactive as it turns into proton.

Currently, it is understood that the one that is holding the protons (nucleus) of the atom together and is said to be preventing the protons from repulsing each other (as they have the same charge) is the pion of the current understanding of the strong nuclear force (Figure 4.4).

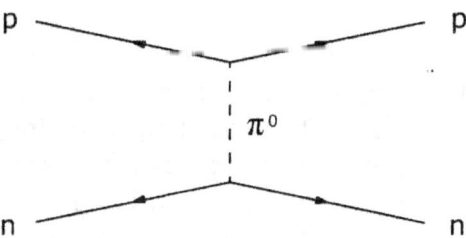

Fig. 4.4. Feynman diagram of the strong nuclear force between proton and neutron mediated by a neutral pion. Time proceeds from left to right.

Pion is classified as a meson, which is composed of two quarks: a quark and an antiquark. The same diagram in Figure 4.4 is shown in Figure 4.5 that supposedly shows how the strong interaction gives rise to the strong nuclear force.

Chapter 4: Theory of Everything: Standard Model with Gravity

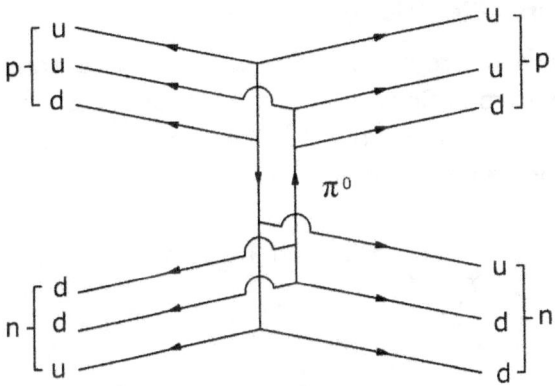

Fig. 4.5. The same diagram in Figure 4.4 showing the quarks of the pion.

It is clearly seen from the quark strong force and the quark weak force to the atom strong force and atom weak force that the energy field is just the magnetic field, which is mediated by the photon. There is a question as to the exchange of the quarks in the proton and neutron or the nucleus. The structure of the quarks (Quark Theory) in the proton and neutron and the structure of the atom (New Model of an Atom) is a much simpler explanation of why the strong interaction of the quarks is passed down as the strong nuclear force of the atom. While it was *observed* that there are positive pion, negative pion, and neutral pion, its role or even its existence should be verified again as to what context they are according to the above discussions. Note that the existence of the positron as an antiparticle of electron (being that the positron is an electron that spins in the opposite direction) can be understood in the context that positron was created through a collision of particles. That is, the observed charge pion should also be understood in the context of collision of particles as the collision of particles can cause a particle to spin in the opposite direction.

Electromagnetism (Electricity)

Electricity (electrical current) is the flow of the electrons in a conductor. The electrons flowing in the wire conductor radiates an electromagnetic field (Figure 4.6). The mediating particle of an electromagnetic field is photon.

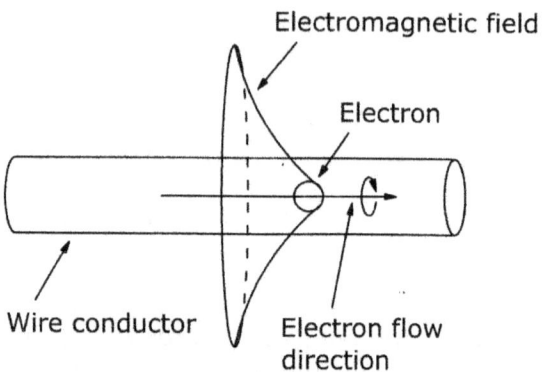

Fig. 4.6. The energy field of the individual electrons flowing in a wire conductor coalesce to extend far outside of the wire to form the electromagnetic field. (Or that the flow of the electrons in a conductor created a "wake" from the energy field of the atoms that extend outside of the conductor.)

The energy field (electromagnetic field) of the moving electrons in a coiled wire conductor (an electromagnet) is practically similar to a magnet because it has a polarity (Figure 4.7).

Chapter 4: Theory of Everything: Standard Model with Gravity

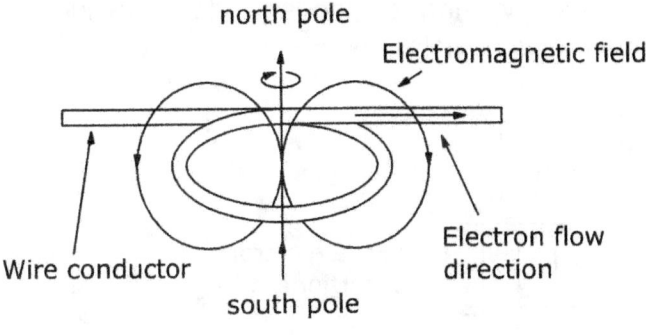

Electromagnet

Fig. 4.7. The same energy field of the electrons flowing in a wire when the wire is coiled produces a similar structure to that of the magnetic field of a magnet.

An electron moving in free space such as in a cathode ray tube or particle accelerator does not produce a magnetic field in the same way as in the electricity (Figure 4.8).

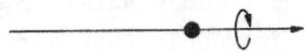

Fig. 4.8. An electron moving in a cathode ray tube or particle accelerator is just a particle that spins in a clockwise direction when viewed in front.

Electron moving from a gap, such as the gap that bends the cathode rays, is said to produce an electric field (Figure 4.9). There is actually no field or electric field that is created but what is thought to be the electric field is actually the electrons themselves moving in free space. The electron is similar to the mediating particle of light (photon), magnetism (photon), and electromagnetism (photon) in that they have the same charge and so the free electrons in a cathode ray tube can be repelled by light, magnetism, electromagnetism, and free electrons (electric field).

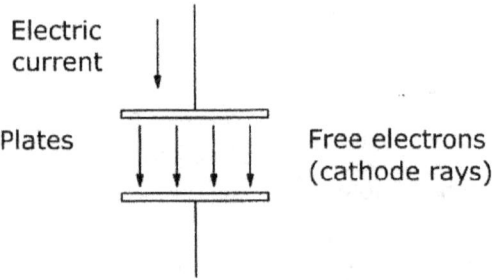

Fig. 4.9. Electrons crossing the gap does not produce an electric field but what is thought as the field are the electrons themselves.

The confusion on the use of the terms magnetic field, electromagnetic field, and electric field in electricity can be explained by the following: Magnetic field comes from the magnet, electromagnetic field comes from the electrons in electricity, and electric field are the streams of free electrons.

Maxwell's unification of magnetism and electricity will be discussed below on section *On Maxwell's Electromagnetic Theory*.

Light

Currently, light is understood to have either a wave or particle property, or both. Light's wave property is currently understood to be an electromagnetic wave. (Electromagnetic wave is a transverse wave of electric field and magnetic field.)

What is commonly called light is the visible light, which is said to be one of the regions in the electromagnetic radiation spectrum. (There is actually no such thing as an electromagnetic wave. The photon of light does not contain all the spectrum of the visible light from violet to red. This is explained

Chapter 4: Theory of Everything: Standard Model with Gravity

in *Appendix A: Theory of Light* where I discussed how I came to understand the nature of light. The complete discussion of the nature of light was discussed in my still unpublished book on my theory of light.)

Light is an emission of an energy particle called photon from an atom such as in a nuclear reaction (nuclear fusion from the Sun) or in the movement of electron in the atom, which then travels outside of the atom or in space.

Light is a charge particle (spinning in a clockwise direction looking in front of it) much like the electron, as such the photon of light is a negative charge particle (Figure 4.10). The photon of light does not radiate an energy field unlike the electron moving in a wire conductor.

Fig. 4.10. Light is an emission of energy particle called photon, which spins in a clockwise direction, hence it is a negative charge particle like the electron.

We can observe this from the cathode ray tube where a magnet's magnetic field deflects the cathode rays, which are free electrons. Considering that light is a photon particle and since the mediating particle of the magnet's magnetic field is photon and it deflects the electron, then photon has the same charge as the electron. Also, since gravity's mediating particle is photon, then it can also deflect the photon of light. (Compare this explanation on gravity "bending" of light to Einstein's theory of general relativity.)

Magnetism

The magnetic field of a magnet is the energy field created by the aligning of the polarity of the magnetic field of the atoms of the magnet, which then coalesce to extend outside of the bounds of the magnet (Figure 4.11).

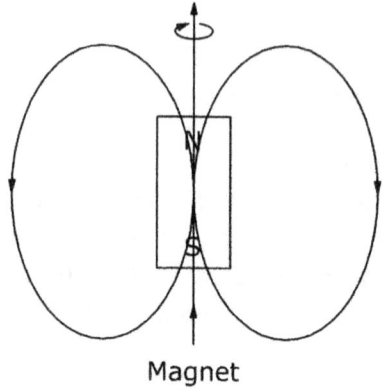

Magnet

Fig 4.11. Magnetic field of a magnet. (The spin shown is the direction of spin of the photon of the magnetic field.)

Note that we can also assume that just as the photon of the magnetic field spins, that magnetic field of a magnet also spins.

The magnetic field of the atom is itself derived from the resultant magnetic field of the proton, neutron, and electron. The source of the magnetic field of the proton and neutron was further derived from the resultant magnetic field of the up quark and down quark. (Although it can also be argued that the source of the magnetic field is from the up quark, which is passed down to the proton.) The mediating particle of the magnetic field is photon.

Gravity

The knowledge of gravity had eluded the great minds of physics such as Galileo, Descartes, Newton, and Einstein. It had eluded the minds as well of the modern physicists so much that they

Chapter 4: Theory of Everything: Standard Model with Gravity

are bewildered as to what it is and say that the standard model cannot explain it.

In *Chapter 3: Quantum Gravity Theory*, I had shown that the source of gravity is the fundamental particles: up quark, down quark, and electron, which are basic dipole magnet energy particles. The magnetic field of the dipole magnet particles: up quark and down quark are passed down to proton and neutron, and with the electron to the atom (refer to Figure 4.1-4.3). The source of gravity is within the atom itself that contains the fundamental particles and thus gravity is within the fundamental particles of the standard model. The "jump" from the fundamental particles to the gravity is created by the large gravitational producing bodies through the dipole gravitational mechanism I had referred to in Chapter 3. My theories on how gravity was created by the gravitational producing bodies was discussed in Chapter 3 on subsection *Theory of How Gravity is Created by the Star, Planet, and Moon* under section *Macro-scale/Macro-universe (Quantum Leap: Dipole Gravitational Body): Planet, Star, Solar System, and Galaxy*.

As I had discussed in Chapter 3, gravity is just a magnetic field or electromagnetic field and its mediating particle is photon. Thus, gravity can be explained by the standard model without the hypothetical graviton. (This brings into question the claim of the string theory on being able to explain gravity and its claim of being the contender for the theory of everything.)

Nuclear Fusion

As I had discussed in my book *Theory of Everything*, when I tried to learn how the elements were created, I studied the creation of the elements from the hydrogen atom to the helium atom taking into account the isotopes of the hydrogen atom. Using my New Model of an Atom, this gave me the idea of the structure of the atom of the elements.

THEORY OF GRAVITY

Nuclear fusion is the process in the fusion of the hydrogen atoms (the source element of the stars) that powers the stars to produce heat in the form of photon of light. Figure 4.12 shows what is called the proton-proton chain.

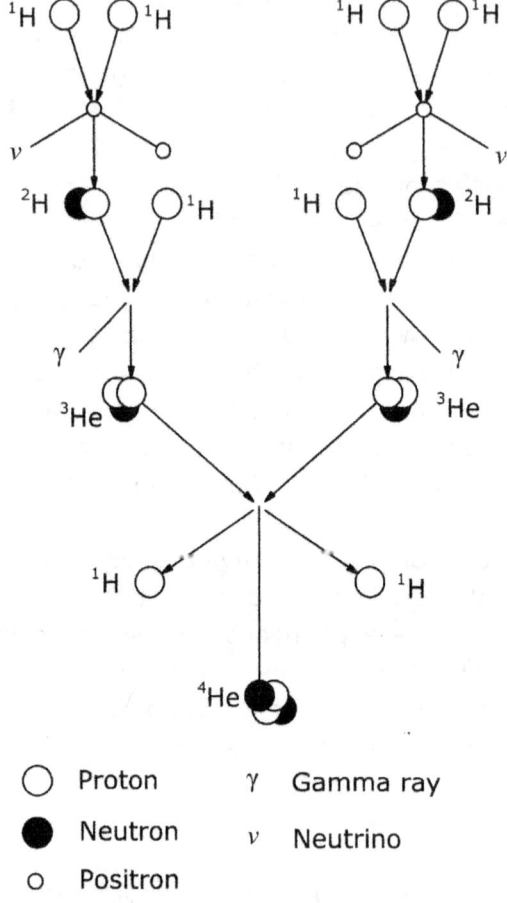

○	Proton	γ	Gamma ray
●	Neutron	ν	Neutrino
○	Positron		

Fig. 4.12. Proton-proton chain of nuclear fusion.

In Chapter 4 of my book *Theory of Everything*, I had shown the structure of elements from the hydrogen atom to its iso-

Chapter 4: Theory of Everything: Standard Model with Gravity

topes of deuterium and tritium to the helium atom, that the nuclear fusion takes a successive process (Figure 4.13).

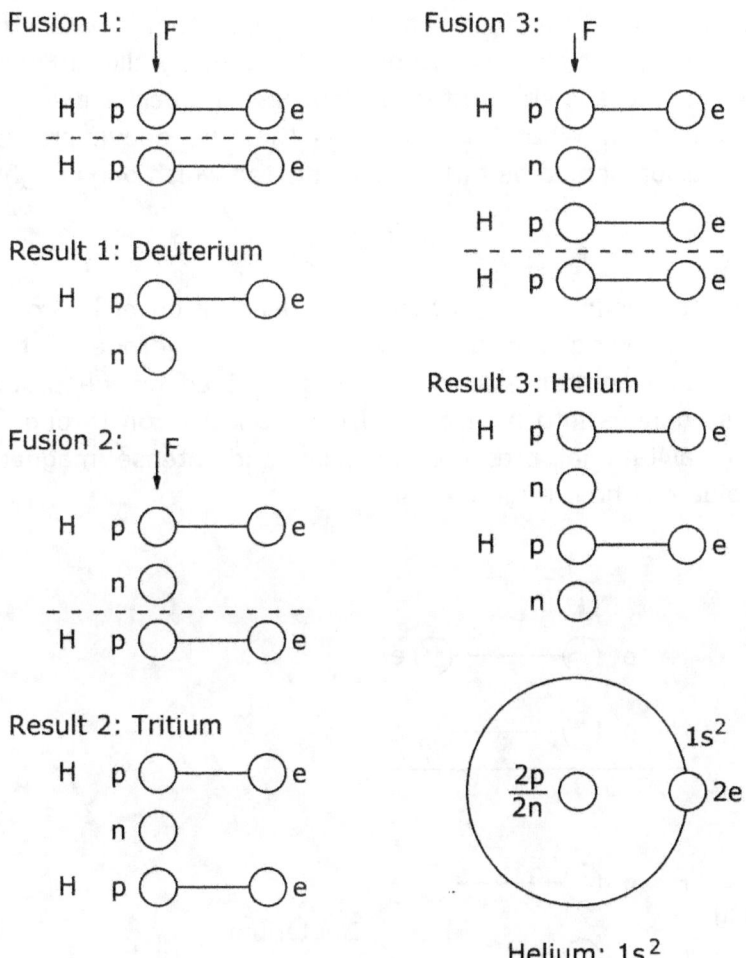

Fig. 4.13. Nuclear fusion of the hydrogen atoms to create the helium atom. (Note: No emission shown in the nuclear fusion.)

When we compare the process of Figure 4.12 and Figure 4.13, we can see that in Figure 4.13 the process requires the creation of the deuterium atom followed by the creation of tritium atom, where it is with the next process of the nuclear fusion that creates the helium atom. It can be observed in Figure 4.12 that the fusion of the tritium atom creates the helium atom and releases two hydrogen atoms, while in Figure 4.13 the process is straightforward successive nuclear fusion of hydrogen atom. What this means is that the above process nuclear fusion should be further investigated which one is right.

* * *

To look deeper into the process of nuclear fusion of hydrogen atoms happening in a star, either a hydrogen atom configuration (proton-electron pair) is retained or the proton nucleus changes into neutron and loses an electron through a force (gravitational pressure or mass and intense magnetic field) aided by heat (Figure 4.14).

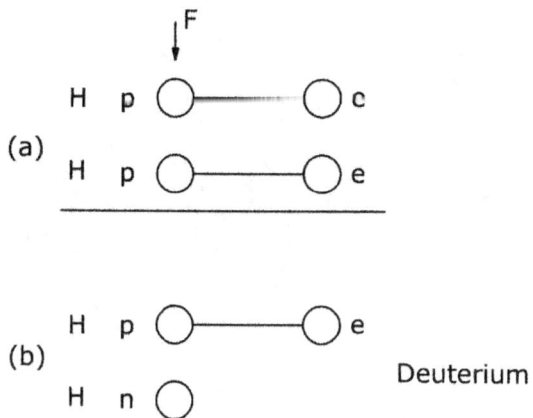

Fig. 4.14. The subatomic process of the changing of the proton to neutron in the nuclear fusion. (F is the force in the form of mass or that

Chapter 4: Theory of Everything: Standard Model with Gravity

Looking deeper into the process nuclear fusion from the transformation of the proton to neutron is the process taking place in the quarks of the proton (Figure 4.15).

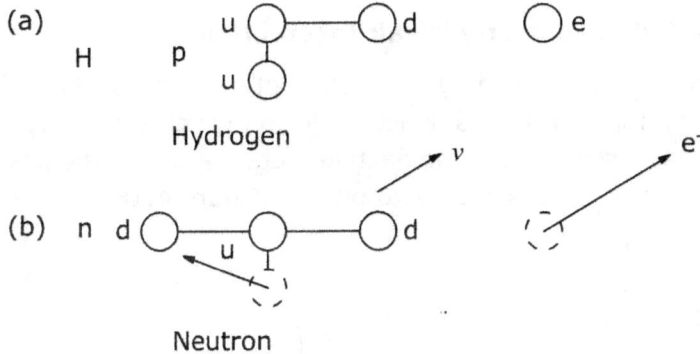

Fig. 4.15. The quarks and electron content of hydrogen atom in the process of nuclear fusion.

About Neutrino and the Source of the Neutrino

It is only through the dipole magnet energy particle with its energy field, which is the cause of the repelling property that we have a solid world. For this reason, neutrino should not be part of the fundamental particles of matter but rather should be placed under the fundamental force of radioactivity (formerly weak interaction). (Neutrino is an emission, which is practically the same as the photon of light, only that neutrino is neutral charge while the photon is negative charge.)

Currently, neutrino is considered as a lepton under the "family" of electron (electron neutrino, muon neutrino, and tau neutrino). We can see in the nuclear fusion in Figures 4.14 and 4.15, and in the radioactivity in Figure 4.17 that if the electron is lost by the proton in the nuclear fusion where the up quark is transformed into a down quark turning the proton into neutron and in the radioactivity where the electron is reacquired and the down quark is transformed into the up quark turning the neutron into proton that the neutrino could only come from the quarks and not the electron. For this reason, the neutrino should not be called an electron's neutrino.

Radioactivity (Formerly Weak Interaction)

The current understanding of weak interaction is that it is responsible for both the radioactive decay, which plays a rule in the nuclear fission. The radioactive beta decay is the transformation of the neutron back to proton (Figure 4.16).

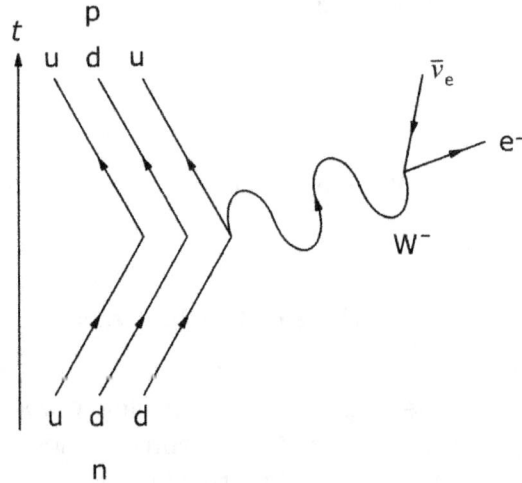

Fig. 4.16. Feynman diagram for the beta-minus decay of the neutron into proton, electron and electron antineutrino through an intermediate heavy W⁻ boson.

For a time I was wavering on replacing the term weak interaction with radioactivity but then I thought about the use of the term strong force to refer to strong interaction and weak force for weak interaction. Strong interaction is the polar magnetic field of the proton that is holding the nucleus together in a pole configuration (Quark Theory and New Model of an Atom). If strong interaction is a force that holds the up quarks

Chapter 4: Theory of Everything: Standard Model with Gravity

in the proton and the proton and neutron in an atom, then it could not be that the weak interaction is also the strong interaction or even related to it. When we think of the weak interaction, it is actually a condition within the neutron that allows it to change back to proton and reacquires an electron. That is, neutron was the transformed proton from the process of nuclear fusion and it is the natural state of neutron to change into proton if there is no force (strong interaction or strong nuclear force exerted by a proton) exerted on it to keep it from changing into proton. It is for this reason that the condition of the changing of the neutron to proton is much better explained by the term radioactivity.

Using the knowledge of the structure of the quarks inside the proton and neutron, the transformation of the neutron to proton gives us a much clearer understanding of the process (Figure 4.17).

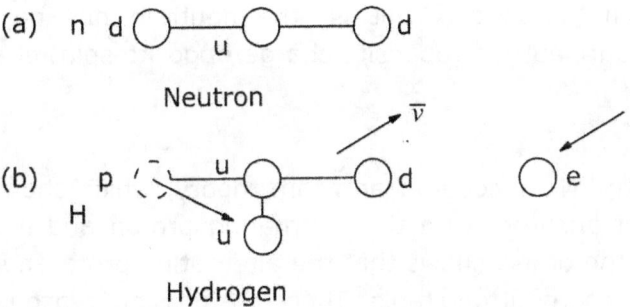

Fig. 4.17. The transformation of neutron to proton.

Figure 4.17 shows that the transformation is straightforward. The down quark turns into an up quark and the proton "grabs" an electron. It also shows that the electron antineutrino does not come from the electron or it is not related to the electron at all but could have only come from the process of the down quark of the neutron turning into the up quark of proton.

THEORY OF GRAVITY

In a way, neutron could be thought of as a spring that is under tension from the process of the pressure of nuclear fusion where the proton was "squeezed" to give up its proton and turn itself into neutron. Neutron is observed to be heavier than proton, which could be explained since its energy field (the resultant energy fields of the up quark and down quarks) is not spinning ($m=E/c^2$) but also because it could have acquired more energy in the process of nuclear fusion released as a neutrino upon changing to proton. (I visualized neutron producing the electron out of its "pocket" but then an atom actually showed there is an excess "somewhere" of electrons since an atom usually has an equal number of proton and neutron being that neutron was a proton before lost the electron.)

There is also a question on the neutrino. Neutrino is a particle that does not spin. That is, a neutral charge particle, which is a particle that does not spin, should have no corresponding opposite. That is, the neutrino has no anti-particle or antineutrino (opposite charge/opposite spin) at all.

* * *

Note that the New Model of an Atom theory, which shows the structure of an atom with the alternating proton and neutron one above the other, shows that the alternating proton-neutron pole makes the neutron stable. There are possible reasons why the neutron becomes unstable or radioactive:

a. The hold of the proton's polar magnetic field on the neutron is weak such as when there are more than enough neutrons for the proton to hold and so the excess neutron starts turning (decaying) into proton.
b. The neutron becomes separated from the atom.

The neutron that becomes unstable could decay by itself with the changing of the down quark to the up quark. The resulting neutron changes to proton and "grabs" an electron (the hydrogen atom configuration, which can be observed in the Periodic Table of a one is to one ratio of proton to electron of a usually stable element where only one electron orbits a proton). (Note that it is not questioned how the up quark turns into the down quark and vice versa.)

On Maxwell's Electromagnetic Theory

Maxwell's electromagnetic theory is said to be the unification of electricity and magnetism. According to Maxwell's electromagnetic theory, a changing electric field (E) produces a changing magnetic field (H), and vice versa.

(Electromagnetic theory of light propagation is said to unify electricity, magnetism, and light. Maxwell discovered that electromagnetic waves travel at the speed of light. As discussed above on subsection *Light* under section *Fundamental Forces*, light is not an electromagnetic wave. More so, an "electromagnetic wave" or an electromagnetic field does not have components of electric field or magnetic field, nor are they perpendicular to each other. The reason why electromagnetic radiation (the so-called waves) from electromagnetic field travels with the speed of light is that the mediating particle of electromagnetic field is photon, which is the same as the photon of light.)

As I had separated the current understanding of electromagnetism into electricity (electron flow or electric current) and magnetism (magnet), it is very important to know what is referred to as "electric field" and "magnetic field" in Maxwell's electromagnetic theory (see discussions above on sections on *Electromagnetism (Electricity)* and *Magnetism*). According to the current understanding of electromagnetic theory, the sources of these fields are the electric charges and electric currents. That is, the electromagnetic theory was based on electricity and the "magnetism" of electromagnetism.

Maxwell's electromagnetic theory is said to have unified electricity and magnetism (magnet) but in truth, electricity and magnetism cannot be "unified" (even if the "magnetism" is an electromagnetism). For a start, electricity came from the flow of electrons while magnetism came from the quarks of the atom. If it is between electricity and electromagnetism, then there is a commonality between them. Electrons and photons of the electromagnetic field (and magnetic field of the magnet) are both negative charge particles; hence, they can repel each other. Also, electricity creates an electromagnetic field and vice versa.

For demonstration purposes, the term electric field usually comes from the electron (electricity) while the magnetic field is shown to be both coming from electricity (electromagnetic field) or the magnet (magnetic field).

To examine the effect of a changing electric field to produce a changing (electro)magnetic field and the changing (electro)magnetic field (or moving magnetic field) to produce a changing electric field, it is best to discuss them one at a time.

A changing electric field from electricity (coming from an alternating current) that produces a changing electromagnetic field can be shown by a wire conductor carrying an electric current and another wire conductor whose electrons will be induced to flow by the aforementioned electric current carrying conductor to produce a changing electromagnetic field (Figure 4.18).

A changing (electro)magnetic field that produces a changing electric field is practically the same as in Figure 4.18.

For the most part, an illustration of a changing magnetic field that produces a changing electric field is shown by a moving magnet that induces an electric current on a wire conductor (Figure 4.19).

Chapter 4: Theory of Everything: Standard Model with Gravity

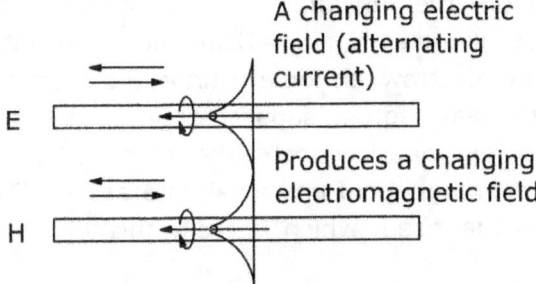

Fig. 4.18. A changing electric field (electromagnetic field) produced by the flow of electron, sweeps the electrons on the other wire conductor inducing the electron to flow and thus creating an electromagnetic field. (This process is called induction.)

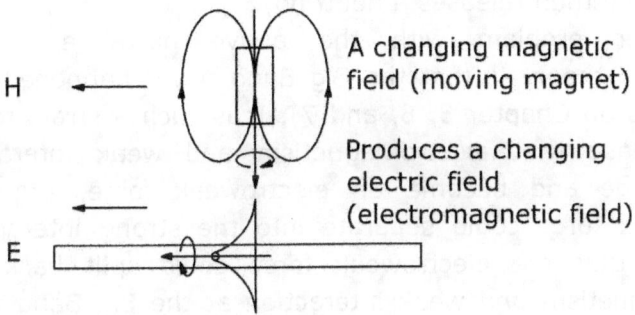

Fig. 4.19. A changing magnetic field shown by a moving magnet sweeps the electrons of a wire conductor that induces the flow of electrons on the wire conductor. (The result is the same as moving the wire conductor on the bar magnet.)

On Electroweak Theory

Electroweak theory is said to be the unification of two of the fundamental forces of nature: electromagnetism and weak interaction. Electromagnetism and weak interaction is said to merge into weak interaction above the unification energy in the order of 100 GeV. Electroweak theory posits that shortly after the Big Bang when the universe was very hot enough (approximately 10^{15}°K), electromagnetism and weak interaction merged into the electroweak force. During the electroweak epoch, the electroweak force separated into the strong interaction. During the quark epoch, the electroweak force separated into electromagnetism and weak interaction. (This is what the physicists theorized, which is really further from the truth.)

From the above discussions on the fundamental forces, there is really no correlation between electromagnetism and weak interaction (radioactivity). Electromagnetism came from the flow of free electron (electricity) while weak interaction (radioactivity) is the changing of the neutron to proton that is happening within the atom (neutron turns into proton, grabs an electron, and then releases a neutrino).

Another problem with the above narrative of the electroweak theory is that the Big Bang never happened (see discussions on Chapter 5, 6, and 7). It is such a strain to our commonsense that electromagnetism and weak interaction could merge and become an electroweak force, then the electroweak force could separate into the strong interaction, and then that the electroweak force could split back into electromagnetism and weak interaction at the Big Bang when electromagnetism, strong interaction, and weak interaction happens only with the existence of the atom.

This goes to show that without the real understanding of what and where these fundamental forces are and their mechanisms, that a wrong theory (electroweak theory and Big Bang theory) supported by mathematics will give us the "answer" even if our premise was wrong. (Why would we think of "unifying" all fundamental forces with the so-called "unification energy" when we even "unify" electricity and magnetism with mathematics that was as least based on the actual interaction of electricity and magnetism?)

(The idea itself of unifying these forces with the so-called unification energy had erroneously pushed us to the idea that a

Chapter 4: Theory of Everything: Standard Model with Gravity

particle accelerator can do it to recreate what is thought to be a moment in the Big Bang.)

Summary: Energy Fields, Emissions, and the Unification of the Fundamental Forces

Bound particles, that is, particles within the atom (up quark, down quark, and electron, which are dipole magnet energy particles; and the proton and neutron within the atom) emit an energy field we called magnetic field (based on the magnet). The magnetic field of the magnet is the extension of the energy fields of the dipole magnet energy particles, which in turn is the proof of structure of the atom. The dipole magnet configuration of the proton, neutron, atom, and gravitational producing bodies inherit their energy fields from the up quark (based from my Quark Theory and New Model of an Atom). The gravitational field came from the dipole magnet energy particles through the structure and mechanism of a body that produces gravity.

Free particles are the photon of light; free electron flowing in a wire conductor (electricity); free proton and free electron in space; and neutrino. Only the free electrons in the wire conductor produces an energy field we called electromagnetic field.

All the fundamental forces of nature are derived from the dipole magnet energy particles. That is, all the fundamental forces of nature exist within the atom and nothing exist outside of it. (We can say that if the Higgs field does not come from the atom then it does not exist.) Based from this, all the fundamental forces can be explained by the standard model.

In the bound particles, the fundamental forces are magnetism, strong interaction, and gravity. In the free particles, the fundamental forces are electromagnetism, light, and radioactivity (formerly weak interaction).

There is a commonality in the fundamental forces of magnetism, electromagnetism, strong interaction, and gravity.

THEORY OF GRAVITY

There is the source of magnetism from the atom and the source of electromagnetism from the movement of electrons (electricity) in a conductor. (Magnetism is often used and confused with electromagnetism as the energy field of the moving electrons in electricity.)

Light (photon) and free electron are similar when the free electron is moving in space; although an electron does not travel very far as it will be captured back by the atom.

Gravity, in the sense of being an energy field, is the same as light in that the mediating particle of gravity is photon and light is a photon.

Magnetic field is often used to mean gravitational field. The gravitational field affects the magnet (magnetic compass) since the mediating particle of gravity and magnetism are both photon.

Strong interaction is practically related to magnetism as they are the result of the energy field of the fundamental particles.

Light and neutrino differ only in that the photon of light spins while the neutrino does not spin. (In fact, we can describe neutrino as a photon that has no charge, as it does not spin.)

To elucidate, these are the "unifying" understanding of magnetism, electricity (electromagnetism), strong interaction, light, and gravity.

Magnetism and electricity (electromagnetism). Magnetism has a different structure of energy field than electromagnetism but both have the photon as their mediating particle.

Magnetism and gravity. Magnetism and gravity have practically the same structure of energy field. Magnetism has its source in the atom just as gravity is a produced by a mechanism by way of the atoms and both have the photon as their mediating particle.

Electromagnetism and light. Electromagnetism is different from light in that the electron travels within the atom and produces an electromagnetic field while light is a photon that travels in space and does not produce an energy field. The

Chapter 4: Theory of Everything: Standard Model with Gravity

understanding of a free electron travelling outside of the atom like in the particle accelerator or cathode ray tube is the same as the photon of light travelling in space. The mediating particle of electromagnetism is photon and light is a photon particle. Both the electron and photon of light has the same charge, hence can repel each other.

Gravity and light. Gravity's photon and light's photon are both practically the same, hence they both repel each other as they are both negative charge particle. (Both are particles and are not waves. This is an important understanding, as it will be used to refute Einstein's theory of general relativity's bending of light by gravity and gravitational waves.)

Based from the above discussions, the fundamental forces are related as follows: the strong interaction, magnetism, electromagnetism, and light are mediated by the photon; and that light (charge particle/spinning) and radioactivity (neutrino/non-spinning) are related to each other as they are both an emission.

Standard Model—Revised

In my book *Theory of Everything* I had published on September 2013, I had revised the standard model as shown in Chapter 1 Figure 1.4. In that standard model, there are only two fundamental particles, the up quark and the down quark, where their so-called "generations" are just their high-energy self. (Ever wonder why the "generations" of the higher-energy up quark and down quark were discovered with the increase in the energy of the machines?) I had kept "open" the "unification" of the fundamental forces that is called for by my theory of everything. This is understandable in the sense that my book *Theory of Everything* was is supposed to be the last book in my three books in physics as my theory of gravity goes first before my theory of everything.

My theory of everything was based on the reality of the dipole magnet energy particles: up quark, the down quark, and the electron, which are the fundamental particles. The fundamental particles are the building blocks of our world, which through their structures (Quark Theory and New Model of an Atom) created the seeming complicity of our world.

In this book, I had finally reorganized the fundamental forces as discussed above into: electromagnetism (electricity), light, quark strong force (strong interaction), quark weak force, atom strong force (strong nuclear force), atom weak force, radioactivity (formerly the weak interaction), and gravity. As I had said in my book *Theory of Everything*, the theory of everything is not about the mathematical unification of the fundamental forces but rather it is about knowing what and where those forces are and how they operate. As I had shown in Figure 4.6, the mediating particle of the energy field of the dipole magnet energy particle such as the up quark, down quark, and electron is the photon. When we follow the formation of the proton and neutron to the creation of the atoms to the creation of the gravitational field producing body such as the star, the planet, and the moon, then the mediating particle of gravity is also the photon.

From the discussions in this chapter, the standard model had been revised into the following:

- Quark strong force (in combination with quark weak force): Strong interaction
- Atom strong force (in combination with atom weak force): Strong nuclear force
- Electricity and magnetism: Separated electromagnetism
- Light: Separated from electromagnetism
- Gravity is added
- Radioactivity: Replaced weak interaction

Chapter 4: Theory of Everything: Standard Model with Gravity

The fundamental forces can divided into the following:

- Matter (atom): Quark strong force and quark weak force (strong interaction), atom strong force and atom weak force (strong nuclear force), electricity, magnetism, gravity
- Emission: Light and radioactivity

Based from the mediating particle, the fundamental forces can be grouped into the following:

- Photon: Quark strong force and quark weak force; atom strong force and atom weak force; magnetism; electromagnetism (electricity); light; and gravity
- Neutrino: Radioactivity

Thus, all the mediating particles of the fundamental forces were replaced with photon except for radioactivity where the neutrino that was once included in the fundamental particles was transferred to the fundamental forces.

Based from these ideas, the standard model could now be fully revised with the complete simplification for which I had included the neutrino to take account for the radioactivity (Figure 4.20).

THEORY OF GRAVITY

Standard Model of Particle Physics
Elementary Particles

	Fermions	Bosons	
Quark	u (c, t)	γ^1	Force Carriers
Quark	d (s, b)	γ^2	Force Carriers
Lepton	e (μ, τ)	v^3	

1. Quark strong force, Quark weak force, Atom strong force, Atom weak force, Electromagnetism, Magnetism, and Gravity
2. Light (photon emission)
3. Radioactivity (neutrino emission)

Fig. 4.20. Revised standard model of the fundamental particles and complete fundamental forces.

Theory of Everything in a Nutshell

The theory of everything is the understanding of the basic building blocks of nature:

- Dipole magnet energy particle: the energy particle (up quark, down quark, and electron), its spin, and its energy field
- Charge theory (spin and direction of spin)
- Mass ($m = E/c^2$)

The discovery of the structure of matter is very important in understanding how matter was formed:

- Quark theory (the up quark and down quark in proton and neutron)
- New Model of an Atom (proton, neutron, and electron)

Chapter 4: Theory of Everything: Standard Model with Gravity

All the so-called "laws of physics" have their basis in the theory of everything, yet the discovery and understanding of the theory of everything will not mean anything without the discovery of the properties, mechanisms, and how things operate that constitutes the laws of physics. Thus, the existing theories that are rigorously proven right and everything I had discovered and written, lays the groundwork for the greater understanding of our world. (Note that the current state of physics where wrong theories such as the Big Bang theory, general relativity theory, special relativity theory, and all their proofs even supported by expensive machines should bring a healthy dose of skepticism.)

Part 2

Theory on the Origin of the Universe

Chapter 5
Cosmology and Theories on the Origin of the Universe

The narrative of cosmology was often written with the Big Bang theory as the established idea and the unassailable final theory on the origin of the universe. (Not surprisingly, fields of particle physics involving the earlier existence of the fundamental particles and the unification of the fundamental forces were made to fit to the Big Bang theory.) While an opposing steady state theory is discussed in books and articles, it is often written off as a footnote of a failed theory. It was a surprise to read an internet article from the website of the American Institute of Physics (AIP) that discussed other ideas:

> https://www.aip.org/history/exhibits/cosmology/index.htm
> https://www.aip.org/history/exhibits/cosmology/ideas/expanding.htm

What I had found from the website was that contrary to the idea of the Big Bang theory that the galaxies are supposed to be receding from us, Vesto Slipher (1875-1969) (discussed below) observed that not all galaxies are receding from us but instead some are coming towards us.

(Not often heard of is the name Halton Arp (1927-2013), an American astronomer who catalogue many examples of interacting and merging galaxies. Arp was a critic of the Big Bang theory.)

Big Bang Theory, Steady State Theory, and Einstein's Cosmology

In the beginning of the twentieth century, astronomers were not sure of the size of our world. In 1916, Harlow Shapley (1885-1975), an American astronomer studying a globular cluster in our galaxy, which is a group of hundreds of thousands of stars, noticed faint blue stars. He reasoned that if they are similar to the bright blue stars near our Sun, then those faint blue stars must be very far to explain their lightness in color. In order to establish distances more conclusively, Shapley based his measurement using the Cepheid, a type of variable star. Shapley was finally able to place correctly our Sun not in the center of our galaxy but on its outskirt. However, he at the same time held the view that our galaxy was the entire universe.

(In 1917, Edwin Hubble (1889-1953) started his work in astronomy and corresponded in later years with Shapley.)

Einstein's Cosmology

In the early twentieth century, the common worldview was that the universe was static.

In 1917, Willem de Sitter (1872-1934), a Dutch astronomer, produced an equation that described a universe that was expanding, that is, a universe with a beginning. Einstein, taking the universe as a whole, used his gravitational field equations to provide just a compact mathematical tool that could describe the general configuration of matter and space. Leading scientists of his time endorsed Einstein's field equation to become a foundation for cosmology.

Chapter 5: Cosmology and Theories on the Origin of the Universe

Einstein and de Sitter were in communication comparing their separate models of cosmology. Willem de Sitter's model could only be stable if it has no matter in it. Einstein's early model on the other hand could not contain matter and be stable. His equations showed that if the universe were static at the start, the gravitational attraction of the matter would make it collapse on itself.

Einstein solved the instability of his model by adding a constant term to his equations. This constant, if not zero, would make his model not collapse under its own gravity. Called "cosmological constant," Einstein admitted that it is only a "hypothetical term" and it is not required by the theory. The cosmological constant was only necessary for the purpose of making possible a quasi-static distribution of matter.

In 1912, Vesto Slipher obtained a spectrogram for the Andromeda nebula indicating that the nebula was approaching our solar system at an amazing speed. Slipher measured the velocities for other spiral nebulae over the next two years. His first few measurements showed approaching nebulae on the south side of our galaxy, while on the opposite side the nebulae are receding. The result made Slipher formed a "drift" hypothesis whereby he thought that it was our galaxy that was moving towards the south nebulae, away from the north nebulae. However, his hypothesis was contradicted by the observations of more spiral galaxies, as receding spiral galaxies were found on the south side as well as on the north side of our galaxy. Nevertheless, Slipher clung to his "drift" hypothesis.

Meanwhile, de Sitter's model of a static universe produced a diminishing frequency of light with increasing distance. By 1921, de Sitter knew of Slipher's measurements of the velocities for the spiral galaxies were the measurements showed only three were approaching. The observation of the three approaching spirals was filed away as the result of large velocities in random directions, superimposed on a much smaller systematic recession.[1]

In 1928, Hubble attended a meeting of the International Astronomical Union held in Holland where he discussed cosmology with de Sitter. Hubble returned to his Mount Wilson Observatory to test de Sitter's theory of shifting frequency. Hubble with his assistant Milton Humason (1891-1972) was looking for a displacement of lines in the spectrum towards the red, in what would later be called the "red shift." What they both found was that the greater the receding velocity of a nebula is, the farther the distance to it. Einstein agreed that the universe is indeed not static.

Back in 1922, Alexander Friedman (1888-1925), a Russian meteorologist and mathematician published a set of mathematical solutions that showed a non-static universe.

In 1927, Georges Lemaitre (1894-1966), a Belgian astrophysicist and a Catholic priest published his model of an expanding universe in a little-read *Annals of the Brussels Scientific Society*. His paper was easily overlooked. When Lemaitre saw a report of the Royal Astronomical Society meeting held in 1930, he wrote to Arthur Eddington (1882-1944), who was his former teacher, reminding him of his 1927 paper.[2] Eddington recognized the value of Lemaitre's paper and shared Lemaitre's paper to de Sitter. De Sitter wrote to Shapley. Einstein confirmed that Lemaitre's work fits well into his general relativity theory. In 1931, de Sitter proclaimed in public Lemaitre's brilliant discovery of the expanding universe.

Creation of the Elements: The Hydrogen-Helium Abundances Issue

In 1946, George Gamow (1904-1968), a Ukrainian-born American physicist pondered the early stage of the expanding universe as a superhot stew of particles and began to calculate the amounts of various chemical elements that might have been created under such conditions. In the expansion and

Chapter 5: Cosmology and Theories on the Origin of the Universe

cooling of a universe from nearly infinite density and temperature, the state of all matter would have been protons, neutrons, and electrons merging in the ocean of high-energy radiation.[3] He thought that elements could be built-up as an atom by capturing neutrons one by one. Gamow succeeded in explaining the cosmic abundances of hydrogen and helium but failed to get a sensible answer for the creations of other elements above helium. (In my book *Theory of Everything*, I had explained using my New Model of an Atom the construction of an atom of an element. Gamow's idea that neutron exist together with proton after the Big Bang is wrong since neutron is the result of the nuclear fusion from the stars where proton is turned into neutron.)

Steady State Theory

In the 1920s, Sir James Janes (1877-1946) was the first to conjecture of a steady state cosmology. He based his hypothesis on a continuous creation of matter in the universe.

In 1948, Fred Hoyle (1915-2001), Thomas Gold (1920–2004), Hermann Bondi (1919–2005), and others revised the idea of the steady state cosmology. The steady state theory asserts that the universe has always expanded at a uniform rate, with no beginning or end, that it will continue to expand and have constant density, and that the distribution of the old and new objects in the universe is basically even. In order to have constant density, matter is created to fill the void left by the galaxies that are receding from each other.

Fred Hoyle was successful in explaining how the rest of the elements after helium were created in the stars interior. (Big Bang theory stipulates that the elements were created from the very start.)

In a radio broadcast program, Hoyle coined the term "big bang" as an argument against Lemaitre's theory that described how the universe was created in an instant from the remote past.

Cosmic Microwave Background Radiation

In a 1948 paper, George Gamow argued that the Big Bang universe would be at first dominated by radiation. As the radiation expanded, it would be created into matter.[4] Ralph Alpher (1921-2007) and Robert Herman (1914-1997) predicted that a remnant of the radiation would remain—a cosmic background radiation permeating all space.

Microwaves were found to be useful for radar and communications. In 1963, Arno Penzias (1933) and Robert Wilson (1914-2000) working for Bell Telephone Laboratory while studying the sky's microwave "noise" realized that they had detected microwave radiations coming from all around the sky. In 1965, Robert Dicke (1916-1997), a physicist working nearby at Princeton University had learned of the measurements of radiation, about 3° Kelvin, correctly interpreted it as the radiation from the Big Bang. Dicke had independently recognized the prediction of Alpher and Herman.

(It is said that by 1965, for all intents and purposes, the debate for the Big Bang theory and the Steady State theory was over—the Big Bang theory won.)

Inflation Theory

In 1979, Alan Guth (born 1947), an American particle physicist proposed a theory that an astonishing rate of "inflation" could have taken place in the first moment of the universe's evolution, then it slowed down.

Chapter 5: Cosmology and Theories on the Origin of the Universe

Dark Matter, Dark Energy, and Einstein's Cosmological Constant

Back in the 1930s, Fritz Zwicky (1898-1974), an astronomer had measured the velocities of galaxies in cluster and announced that the cluster should fly apart as the gravitational pull of the visible matter was not enough to hold the fast-moving galaxies together.

In 1970s, Vera Rubin (born 1928) and Kent Ford (born 1931) was surprised when they measured the rotations of individual galaxies and found out the same problem: the outer stars that were orbiting so fast should fly off if nothing is holding them but the gravitational pull of the visible stars. To reconcile this observation with the law of gravity, scientists proposed that there is matter that we cannot see. They called this matter as "dark matter." As such, dark matter is thought to be as-yet undetected matter that provides galaxies with enough mass to prevent the speed of their rotation from pulling them apart.

Early in the twenty-first century, observation of supernovae seems to show that there must be something causing the universe to actually accelerate. Scientists called this "thing" that drives this acceleration as "dark energy." With these new revelations, scientists had retrieved Einstein's cosmological constant and added it back to the theory of general relativity equations. The cosmological constant is now the leading idea to account for the "dark energy."

* * *

The Nobel Prize in Physics in 2011 was shared by awarding half to Saul Perlmutter (born 1959) and the other half jointly to Brian P. Schmidt (born 1967) and Adam G. Riess (born 1969) "for the discovery of the accelerating expansion of the Universe through observations of distant supernovae."

The Universe as Observed: Cosmic Background Radiation, Constant Temperature, and Filament

Gamow's idea about the presence of radiation in the early universe of the Big Bang, the prediction of Alpher and Herman of the remnant of radiation from the Big Bang, and the discovery of Robert Dicke of the constant temperature of the universe led to the acceptance of the idea of the cosmic background radiation.

Early measurements of the CBR revealed that it was uniform to less than one part in a thousand. Theory showed that CBR should not be uniform as there should be tiny, random fluctuations in the temperature of the CBR. That is, since the CBR is an early snapshot of the universe and if the current observation of the universe is "lumpy" with galaxy clusters and superclusters, then surely the early universe was not completely smooth.

Since fluctuations in the CBR were too small to detect with ground-based radio telescopes, there was a need to launched satellites in space to make a more accurate measurements.

In 1989, NASA (National Acronautics and Space Administration), a US government agency, launched the satellite COBE (Cosmic Background Explorer) to measure the microwave radiation. In 1992, COBE came back with a result that showed that the random fluctuations in temperature over the sky are just one part in 100,000.

In 2001, NASA launched a more advanced satellite called WMAP (Wilkinson Microwave Anisotropy Probe) designed to measure the CBR with higher sensitivity and angular resolution, which produced a more detailed full-sky map.

In 2009, ESA (European Space Agency) launched Planck satellite to observed anisotropies of the CBR in the microwave and infrared frequencies with higher sensitivity and small angular resolution. Planck complements and improves upon the observation of WMAP of NASA.

Chapter 5: Cosmology and Theories on the Origin of the Universe

The result of these advancements in satellites is that the view of CBR had been much more detailed. What the images showed was that even with some "cold spots," the universe is still *evenly spread*, which accounts for the constant temperature of the areas of our universe.

These images, though, showed only a two-dimensional image of our universe. (Our universe is also viewed in a three-dimensional view where galaxy filaments were observed. Galaxy filaments are massive, thread-like formations.)

The Mistakes That Are Supporting the Big Bang Theory

Most of the current problems in cosmology in the understanding of the universe and its origin were brought about by the Big Bang theory. There were three mistakes that are supporting the Big Bang theory: the observed expansion of the universe, the observed radiation in the universe, and the observed acceleration of an expanding universe.

Lemaitre theorized of the universe in a Big Bang and Hubble found the "proof" of an expanding universe. Gamow theorized of the dominance of radiation in the early universe and Alpher, Herman, and Dicke found the "remnants" of radiation. From the expanding universe Perlmutter, Schmidt, and Riess had "observed" an accelerating universe.

As it is, a theory was forwarded, a proof was discovered, and the effort was awarded with a prize as a validation for the work. In Chapter 6, I am offering a new theory that could overturn the Big Bang theory. In Chapter 7, I am overthrowing the observations and the proofs of the Big Bang theory.

Chapter 6
Hydrogen Origin Theory of the Universe

Hydrogen Origin Theory (HOT) of the Universe

Our universe is practically infinite and had existed for a long time. In the beginning, our universe was cold and dark and there were only the hydrogen atoms. As time goes by, the hydrogen atoms attract one another (through their magnetic fields) forming from hydrogen gas into watery or solid ice masses.

Early Nebulae

The earliest nebulae were only made up of hydrogen atoms and the liquid or solid hydrogen masses.

A part of a nebula with heavy concentration of hydrogen atoms will create the first star. (Later a nebula will be "dusted" by the solid particles of star that had supernova.) If a giant (spinning) star is formed, it will gather the parts of the nebula to start it spinning into a start of a galaxy.

Star

When some hydrogen masses of a nebula had coalesced where its mass is large enough (as mass produces also strong magnetic field) to create nuclear fusion, the process ignites the birth of a star.

The spin of the quarks and the atom is translated to the star, which make the star spin and gather more materials from nebula. The star itself will produce heavier element like helium in the process of nuclear fusion.

Abundance of Hydrogen and Helium

Hydrogen and helium in the universe account by mass to about 73% and 25%, respectively.

The universe started with hydrogen atoms and that its evolution is the creation of the star, which is powered by nuclear fusion of the hydrogen atoms that creates helium and light. It is but natural that the most abundant elements in our universe are the hydrogen and helium.

Galaxy

As a nebula evolves, stars are created. When a very massive star is created strategically in the nebula, it will spin and gather parts of the nebula into spherical shape, which is the beginning of a galaxy. As this spherical galaxy spins, centrifugal force takes into effect flattening the galaxy into an elliptical galaxy. As the galaxy evolves and ages, the materials in its parts also form stars and star systems (solar system). (It is very possible that solar system is created in the nebula.) A nebula may not always form a galaxy, which will make it an irregular nebula.

Supernova

Supernova happens when the materials of the stars turned into the heavy elements. This is known to be triggered usually when a star produces iron. Supernova makes all of the elements that make the life in our universe possible. The exploded supernova becomes the added material for a nebula, stars, and galaxies.

It is interesting to note that a star that will supernova should be very massive to create those elements at the end of our periodic table.

Galaxy Clusters, Superclusters, and Filaments

Galaxy clusters are structures in the universe where galaxies group themselves through the dynamics of their gravitational fields (polar gravity and field spin gravity). Galaxies could still create a much greater structure called superclusters. However, the greatest structure of the universe is called filaments, which are massive, thread-like formations of the bodies of our universe. (Although the dynamics of field spin gravity and polar gravity shapes the clusters and filaments, overall, the polar gravity is primarily the one that shapes these thread-like structures of the filament.)

Cosmic Background Radiation (CBR)

Our universe is full of stars, which is the common source of its light. Cosmic background radiation is caused by the light from the stars. The energy regions (the so-called electromagnetic radiation spectrum) of the photon (from the highest to the lowest) are: gamma rays, x-rays, ultraviolet rays, visible light, infrared rays, microwave, and radio waves. What was observed as the cosmic *microwave* background radiation was just the energy of the photon at the microwave range. In fact, the galaxies and parts of the universe was photograph in various spectrums such as in the x-rays or infrared rays to see its details or what is hidden by the "dust." The so-called "cosmic microwave background radiation" is not the "embers" of the Big Bang at all but the photons from the stars that had been reduced to that energy.

Constant Temperature of the Universe

If our universe is practically homogeneous as it can be seen from the images taken of the cosmic background radiation, it is but natural that our universe will have a constant temperature. This temperature we should say was how the stars had warmed up our universe from its early dark and cold beginning.

Age of the Universe

The age of the universe according to the Big Bang theory is estimated according to the time elapsed since the Big Bang. The problem with this method is that we do not know where the center of the universe is. For example using Figure 6.1: If point A is the center of the universe where the Big Bang started and Earth is at point B and we are looking at the star at point C as the age of the universe, then we only measured a quarter of the age of the universe, which is from point A to point D, as we are limited by the reach of our instrument. Now supposed our instrument can perceived the distance of point E, then we have measured the age of the universe by over 50%.

For the rate of the expansion of the Big Bang, how do we know where is its periphery if we do not even know where is its center? The point is, in the Big Bang theory, we do not know where the center of the universe is and so we have no way of knowing the age of the universe. Now, if the Big Bang really happened, we are supposed to see a large area in the night sky where there were no stars.

* * *

If we will accept my Hydrogen Origin Theory (HOT) of the Universe, we have to determine what constitutes the beginning of the universe. Our universe existed in eons past before the birth of the first star. If we use the beginning of the universe when the first light had started, then we have to find this star

Chapter 6: Hydrogen Origin Theory of the Universe

and possibly, it had already burst into supernova. The gathering of hydrogen into nebula and the formation of galaxies is being repeated throughout the universe. Large size galaxies may take longer to evolve than the smaller ones. Larger stars that developed first burn its hydrogen fuel faster than smaller ones. Still, at the very least, we know that galaxies evolved from the nebula and we could use them to estimate the age of the universe. Maybe a comparison of the same size galaxy could be used at best.

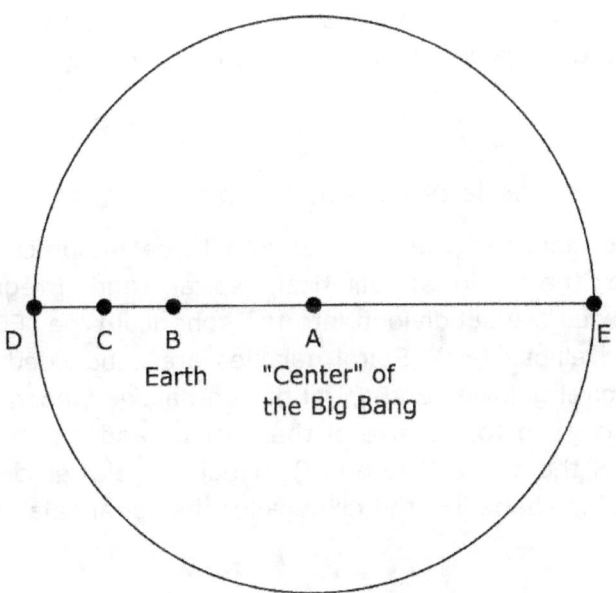

Fig. 6.1. Given that we do not know the center of the universe, the determination of the age of the universe according to Big Bang is not conclusive.

The bottom line is that the Hydrogen Origin Theory (HOT) of the Universe shows that our universe has a practical beginning compared to the improbable beginning of the universe of the Big Bang. It is for this matter that most of the proofs that are used to support the Big Bang can be explained by my theory and that those other theories related to the Big Bang had to be "massaged" to fit it. One of these is the explanation for the abundance of hydrogen and helium. Another is the observation that the universe is expanding even when galaxies are colliding from each other and just as some galaxies are moving away from us, some are coming "towards" us. (Coming towards us or moving away from us does not mean that it will collide on us or it is directly moving away from us just as bodies that are moving parallel with us will be observed to come towards us or moving away from us. On the other hand, Andromeda galaxy is observed to be coming directly towards us.)

Galaxy Classification

Edwin Hubble classified the galaxies into three major classes according to their forms: elliptical, spiral, and irregular. Elliptical galaxies are subdivided into the spherical type (E0) to the elongated ellipse (E7). Spiral galaxies are subdivided into the normal spiral galaxies and spiral barred galaxies, each with types and according to the size of the nucleus and the degree of openness of the arms (Figure 6.2). Irregular galaxies do not have discernable shape like the elliptical or the spiral galaxies.

* * *

What the Hubble's classification of galaxies showed is actually the evolution of a galaxy, which could be used to determine the approximately age of the galaxy. (In theory, it can be used to determine the approximate age of the observable universe with a recognized oldest galaxy.)

From the Hubble's classification of the galaxies: the E0 evolves to E7, the Sa to Sc, and SBa to SBc. However, there is a misunderstanding in the classification of the galaxies as a linear evolution from E0 to S0/SB0. That is, the galaxy does not

Chapter 6: Hydrogen Origin Theory of the Universe

have to progress from E0 to E7 to S0 for the arms to deploy at Sa. It will be observed that even at E0 to E3, the arms could deploy. The reason for this is that the size of the galaxy should be taken into account since the larger the size of the galaxy, there is a chance that it will reach its maximum limit of its gravitational field (field spin gravity) and earlier release its arms. (The release of arms is dependent on the maximum limit of the field spin gravity and its arms are dependent on the polar gravity of its materials.)

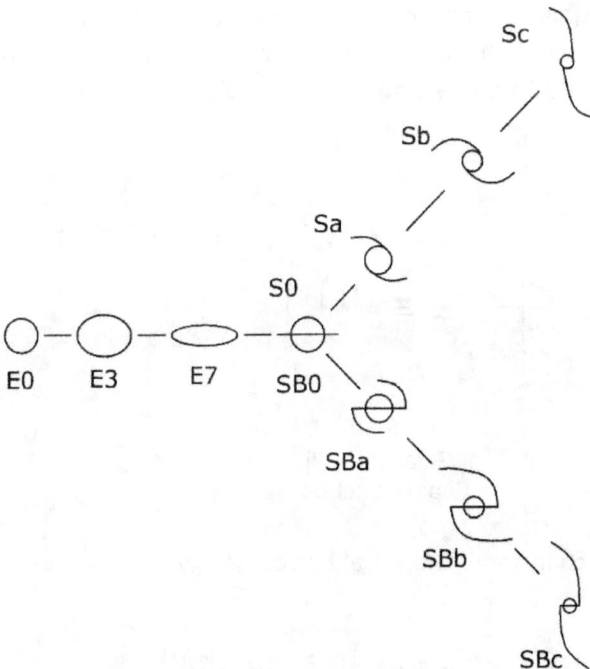

Fig. 6.2. Hubble's classification of the galaxies.

Spiral barred or the physical reality of a "bar" in a galaxy is just not practical due to the immense size of a galaxy (in light years across) and its spin. As the galaxy has a polarity, it will repulse or attract other galaxies as it is "falling" into somewhere in space or aligning within the cluster or the larger filament. Thus, the spiral barred galaxies are actually to some extent an elliptical galaxy that is reorienting itself in attraction or repulsion to another galaxy. Notice that the early unfolding of the spiral arm can be observed to be created directly on the opposite sides of the galaxy possibly simultaneously, which is either through symmetry or through the agent of the field spin gravity. The spiral barred galaxy's "bar" is actually the side view of the galaxy where its arms spin with its axis perpendicular to the axis of the galaxy (Figure 6.3).

Spiral bar galaxies can only be called on how it looks like, although I wonder if the images that are often shown in books were selected only to show how a galaxy with a "bar" looks like.

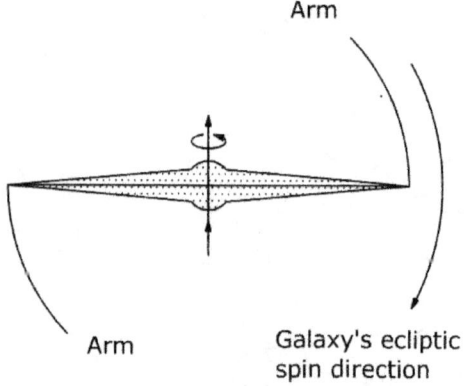

Fig. 6.3. The mechanism of the spiral barred galaxy.

HOT Universe in a Nutshell

The Hydrogen Origin Theory (HOT) of the Universe is a very simple theory on the origin of the universe in that the universe is an infinite with unknown dimension that started with the hydrogen atoms, dark, cold, and later became watery. Through the coalescing of the hydrogen atoms nebulae were started, then the stars, then the supernova of the early stars, seeding

Chapter 6: Hydrogen Origin Theory of the Universe

the universe with solid masses (elements) that would later become the parts of the stars, planets, moons, meteors, asteroids, and dust.

Part 3

Overthrowing Theories and Ideas

Chapter 7

Overthrowing the Cosmic Background Radiation, the Observation of an Expanding and Accelerating Universe, and the Ideas of Dark Matter and Dark Energy

CBR as the Remnant of the Big Bang

The cosmic background radiation (CBR) was said to be the remnant of the Big Bang. In Chapter 5, I had discussed in subsection *Cosmic Microwave Background Radiation* on section *Big Bang Theory* that George Gamow was the first one to offer the idea of radiation as the remnant of the Big Bang.

At this stage, I had already said that the Big Bang did not happen at all. To say that the observed "cosmic background radiation" is the remnant of the Big Bang belies the fact that everywhere in the universe are the different energy photons from far away stars whose energies had been reduced to the microwave range—the range commonly associated with the CBR. That is, radiation is actually the photon that was emitted by a star originally as a high-energy photon in the energy of the gamma ray range (high frequency/short wavelength), which through the distance it travelled was reduced to the observed microwave range (low frequency/long wavelength). In fact, the study of the structure of the universe (filaments), stars, nebulae, and galaxies aside from those visible through the ordinary telescope (that is, the world as we view it in the visible light spectrum) uses the different range of the energy spectrum of the photon of light like the gamma rays, x-rays, ultraviolet rays, infrared rays, and the microwaves. These are the photons emitted by the stars that we can say whose energy will lessen with the distance (time) as it travels away from its source.

Commonsense and observation will tell us that if there was an explosion in the creation of the universe, that part of the observable universe should have less material in it of which a hole should be observed and that it should not create an even temperature that is observed in the current age of the universe.

Expanding and Accelerating Universe: Universe's Filament

Hubble's observation of an expanding universe supported the Big Bang theory, which made the theory the most dominant theory on the origin of our universe. In Chapter 5, I had discussed how Vesto Slipher's observations were ignored that galaxies are receding and approaching on the south side and north side of our galaxy. One of the most ignored observations is that if our universe is expanding like a balloon, there would be no collision of galaxies and as it happens, galaxies that collide with each other are a common occurrence.

Our universe is vast and any observation that could be used to support an expanding universe could easily be accepted even if it ignores some of the contradictory observations.

In Chapter 6, I had overthrown the Big Bang theory with my Hydrogen Origin Theory (HOT) of the Universe. In the HOT Universe model, large bodies such as the stars and galaxies with their gravitational field make the dynamic evolution of our

Chapter 7: Overthrowing the Cosmic Background Radiation...

universe through attraction and repulsion. The resulting evolution of our universe created the filaments, which are a thread-like or web-like structure of the universe. Through the structure of the filaments, we can understand why our universe is thought to expand and even accelerate.

Expanding Universe

The filament is the defining structure of our universe and the dominant force that is shaping it is the polar gravity component of the gravitational field. In the web-like structure of the filament, a web is the alignment of the bodies in it such as the stars and galaxies. Figure 7.1 shows why we observe our universe is expanding or why Vesto Slipher had observed that some of the galaxies are receding and approaching our galaxy on opposite sides.

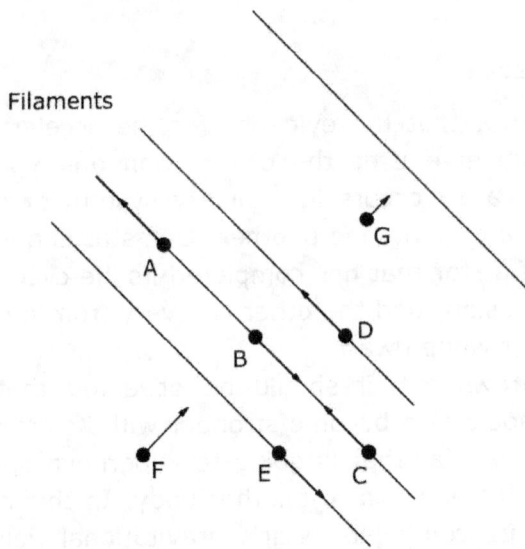

Fig. 7.1. The structure of the filament to illustrate the observed receding and approaching galaxies to our Milky Way galaxy.

From Figure 7.1: A, B, C, D, and E are galaxies that are in a filament thread, and F and G are galaxies that are not in a filament. A is our Milky Way galaxy, B galaxy is receding from us, and C galaxy is approaching us. C galaxy may be attracted to B galaxy or C galaxy is being repulsed by a nearby galaxy, setting it into collision with B galaxy. D galaxy and E galaxy, which are not directly aligned to the filament of our Milky Way galaxy, will be observed that D is moving closer and E is moving away from our galaxy. F galaxy, which is close to our filament, will be observed to be moving closer towards us, possibly to align with our filament. G galaxy, which is not in our filament, will be observed to be moving away from us, merging to another filament.

Thus, what we observed as the expansion of the universe is actually the ordering of the bodies of the universe into the filament structure of the universe governed by gravity.

Accelerating Universe

It is by no coincidence that the evidence for the accelerating expansion of the universe uses the observation on Type 1a supernova. Supernova 1a occurs in a binary system of stars wherein two stars are orbiting one another: One star is a white dwarf (a remnant of a star that has completed its life cycle and has ceased nuclear fusion) and the other can vary from a giant star to even a smaller white dwarf.

When a star grows old, it should be observed that its gravitational field should also become stronger with its creation of heavier elements. A star that is going to supernova should be accelerating due to repulsion to another body. In this case, the star could use its companion star's gravitational field to propel itself into acceleration when it explodes. If it is not accelerating due to repulsion to another body, then its explosion practically accelerates it.

Chapter 7: Overthrowing the Cosmic Background Radiation...

With the billions of stars and some that will be observed to have exploded, it will not be a surprise that some will be observed to recede or approach us from our view. That is, astronomers could refute the expansion of the universe themselves by finding evidence that some supernova 1a stars are receding or approaching from us on one side.

Dark Matter and Dark Energy

In Chapter 5, I had discussed that in the 1930s Fritz Zwicky observed that with the velocities of the galaxies in clusters, the clusters should fly apart since the gravitational pull of the visible matter was not enough to hold these fast moving galaxies.

In the 1970s, Vera Rubin and Kent Ford had observed that the rotations of the individual galaxies showed that the outer stars were orbiting so fast that they should fly off if nothing is holding them but the gravitational pull of the visible stars. They called this invisible force that is holding the galaxies together as dark matter.

Dark energy, on the other hand, pushes everything apart and making the universe expand in which current observation also states that the universe's expansion is accelerating.

(The subject of dark matter and dark energy could have been explained easily after the discussions on *Chapter 2: General Theory of Gravity* and *Chapter 3: Quantum Gravity Theory* but it is best to be discussed it here in the context of the discussions on the theory of the origin of the universe and the universe as it is observed.)

The explanation for the dark matter and dark energy involves understanding the dynamics of a galaxy, particularly those galaxies that had flattened already such as the elliptical and spiral galaxies.

THEORY OF GRAVITY

Breaking Down the Dark Matter

The universe is said to be made up of 26.8% dark matter that does not emit light or interact with regular matter other than through gravity. Based on the understanding in Chapter 2 from the quarks and electron to the proton and to the atom, which are all dipole magnet energy particle, we know that if dark matter is matter, then it has to interact with the "other" matter. If dark matter interacts with gravity, then it has to be at least solid as matter does and are observable or discernable. That is, the explanation for the dark matter does not make sense.

Dark Matter: Galaxy

The original observation for the dark matter was regarding the galaxies. Rightfully so, this is where we can explain what is this so-called "dark matter." (How it was determined that "dark matter" is 26% of the universe and why "nothing" can be measured is beyond me.)

A nebula that formed a central star may eventually spin into a spherical galaxy. As the spherical galaxy increases its spin, it collapses due to centrifugal force into a flatter elliptical galaxy. The form and structure of a galaxy is governed by its gravitational field composed of the field spin gravity and polar gravity (Figure 7.2).

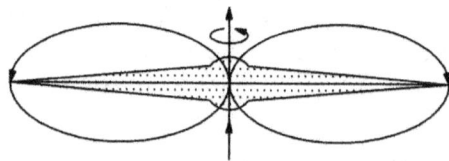

Fig. 7.2. Gravitational field of a galaxy. (Spin shown is the direction of the spin of the galaxy.)

(Vera Rubin and Kent Force observed that the rotation of the individual galaxies showed that outer stars were orbiting so

Chapter 7: Overthrowing the Cosmic Background Radiation...

fast that they should fly off if nothing is holding them but the gravitational pull of the visible stars.)

The material of the galaxy, especially the elliptical galaxy, is controlled by the field spin gravity up to its maximum reach. Whether the field spin gravity still control the periphery of the galaxy or has reached its limits and thus it has no more control of its material on the periphery, its outer material is not only being hurtled by the field spin gravity but also by the possible attraction and repulsion due to polar gravity. So, the cause of why the outer materials of the galaxy does not fly off could only due to the polar gravity. (The astrophysicists may verify this by observing the polar orientations of the stars.)

* * *

As the galaxy evolves, its field spin gravity will reach its maximum reach, losing control of its outer materials, and thus the outer material breaking off to become its arms. The possible cause of the "deployment" of the arms could be repulsion of its outer material stars. It will be observed that the arms will break off at its opposite ends. How this is "communicated" even at such distance of the opposite ends of the galaxy is amazing. From this point, the arm is primarily governed by the polar gravity, as it is the force that is holding the shape of the arm (Figure 7.3).

The proof of the mechanics of the polar gravity can be observed from the spiral Whirlpool galaxy (M51) where its arm is pulling the material of a neighboring smaller galaxy (Figure 7.4). That is, this is the proof that the outer material of the galaxy is governed by the polar gravity.

Note that eventually for the spiral galaxy, the arms of the galaxy will unwind from the center until possibly the arms just becomes streams.

THEORY OF GRAVITY

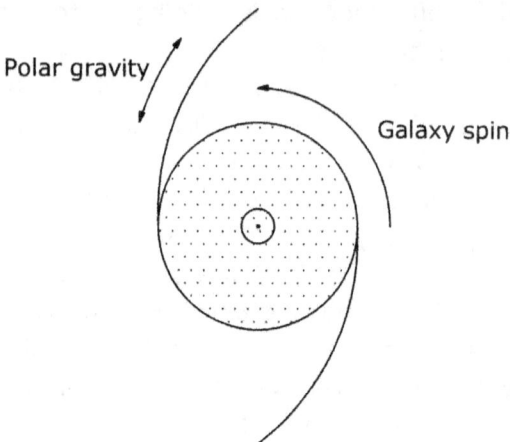

Fig. 7.3. As the field spin gravity created by the center of gravity is created, the arms released the spiral galaxy is primarily governed by the polar gravity.

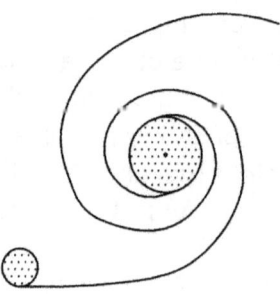

Fig. 7.4. Polar gravity of the arm of Whirlpool galaxy in action, pulling the materials off the smaller galaxy.

Dark Energy: Filament of the Universe

Dark energy is said to be an unknown form of energy that permeates all of space and tends to accelerate the expansion of the universe. The concept of dark energy was derived from the observation of an accelerating expansion of the universe in the

Chapter 7: Overthrowing the Cosmic Background Radiation...

observation of the Type 1a supernova and then applied to the whole universe.

From what was discussed in the section on *Expanding Universe* and on the *Accelerating Universe*, there is actually no expansion and acceleration of the universe—there is only the ordering of these bodies in forming the filament structure of the universe.

The mechanics of the movements of the stars and galaxies is through their attraction and repulsion with each other much like the charge particles, which is due to their spin (Figure 7.5).

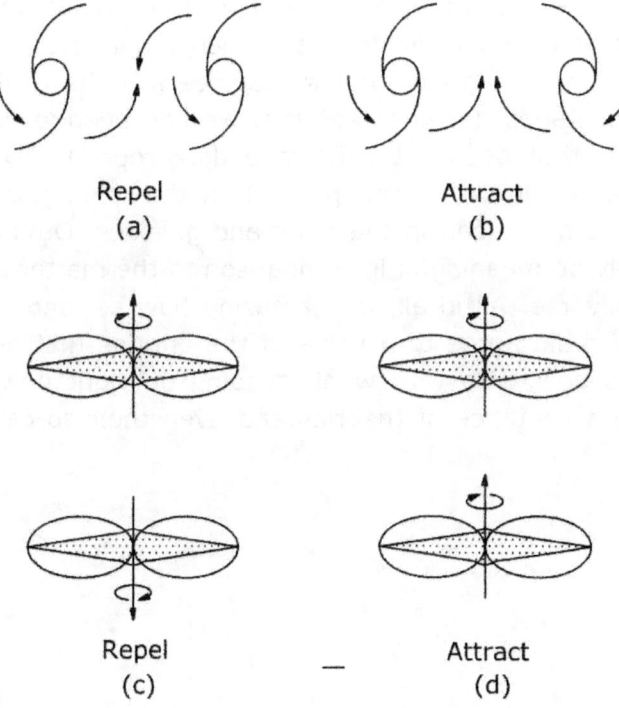

Fig. 7.5. Attraction and repulsion of the galaxies, especially those that are particularly aligned.

Of course, galaxies moving at enormous velocities could collide at any position and attraction and repulsion may not mean anything to these enormous bodies.

In our universe, there is no above or below and so bodies could be falling up, falling down, or falling to the sides and become "entangled" in the structure of the filament.

Percentage of Dark Matter, Dark Energy, and Ordinary Matter

It is said that our universe constitutes 26.8% of dark matter, 68.3% of dark energy, and 4.9% of ordinary matter. We know at this point that dark matter and dark energy does not exist and are just the actions gravity but we can see why these were the given percentages. (We can see that we had been looking for a black cat that does not exist in a dark room.) "Dark matter" can be only in the galaxy, particularly the spiral galaxy. "Dark matter" can be seen in the stars and galaxies. Ordinary matter can only be meaningful in comparison to the matter and void of our universe. All in all, it is amazing how astronomers and physicists could agree to a value of the entities that does not exist. This goes also with what is going on right now in physics in the acceptance of theories and even their so-called proofs.

Chapter 8
Overthrowing Einstein's Relativity Theories

Note: The structure of this chapter is indicated by the three dashes.

Ether
    Aristotle
    René Descartes
    Isaac Newton
    Christiaan Huygens
    Thomas Young
    Augustin-Jean Fresnel
    James Clerk Maxwell
    Michelson and Morley Experiment
    FitzGerald and Lorentz: Contraction Hypothesis
    Ether and Light

Theory of Special Relativity
Theory of General Relativity

How to Overthrow the Theories of Special Relativity and General Relativity
Inertia and Uniform Motion
Disputation of Galileo's Relativity

THEORY OF GRAVITY

Disputation of Special Relativity (Constant Velocity)
    Disputation of the Principle of Relativity
    Disputation of the Principle of the Constancy of the Speed of Light: Ether
    On Electromagnetism and Relativity
Refuting the Effects of Special Relativity
    On Time Dilation
    On Length Contraction

Overthrowing Einstein's Understanding of Gravity
    Einstein's Arguments
    Overthrowing Einstein's Arguments
Disputation of General Relativity (Constant Acceleration)
    Structure of Gravity of General Relativity
    Mechanism of Gravity of General Relativity
Refuting the Proofs of General Relativity
    Gravitational Lensing
    Orbit of Mercury
    General Relativity's Spacetime Curvature and Elliptical Orbit of the Planets
    Black Hole
    Gravitational Waves
    Gravitational Redshift
Commentary

I found that the best resource about Einstein's ideas on ether, inertia, special relativity, and general relativity was the book by Walter Isaacson titled *Einstein: His Life and the Universe* published in 2007. It is unlike other books I had read about Einstein in that it presents almost everything about how Einstein came up with his relativity theories and his ideas about it—even without mathematics.

Chapter 8: Overthrowing Einstein's Relativity Theories

Note: I had laid the foundation of my theory of gravity in Chapter 2 especially about the subject of inertia that I will use to dismantle Einstein's special theory of relativity and general theory of relativity.

Ether

The existence of ether had been entertained by the ancient Greeks and it had survived for centuries through different reincarnations. Various properties of ether were conceived to conform to a particular theory or idea.

Ether was said to have been "blown away" by Einstein's special relativity, yet his general relativity was deceptively based on it with his "fabric of spacetime." It is for this reason that to the present times the existence of ether had never been resolved. To the present times, physicists may not know that their theories (general relativity and Higgs field) were based on ether but without any other theories to replace their cherished theories, the status quo is being maintained.

Aristotle

Aristotle (383-322 BC), a Greek philosopher just like his teacher Plato adopted a Greek philosophy that the universe was made up of the four elements: fire, air, water, and earth. He also adopted his teacher's idea of a fifth element called aether (also spelled ether; later Latinized by the medieval alchemists as quintessence), which is a subtle medium. The four elements occupied the terrestrial space and have a natural place: the earth at the center of the universe, then water, then air, and then fire. When these elements are out of place they tend to go back to their natural place: bodies sink in the water, air bubbles rise up from the water, rain falls, and flame rise up in the air. Ether occupied the celestial spheres and heavenly bodies, and it has a perpetual circular motion.

René Descartes

Aristotelian philosophy still holds sway during the time of Descartes. Descartes (1596–1650) believed in the existence of the ether and hypothesized an ether that is a very dense medium made up of very small particles that pervaded the space (the Higgs boson of the Higgs field should come to mind, which will be discussed in Chapter 9) and that ether transmitted forces from one object to another by collisions of these particles. Descartes believed that the planets are moving in their orbit through a cosmological-sized vortex motion of the ether (see Chapter 1 Figure 1.2). In his vortex theory, the rotating motion of the Sun's ether-vortex makes the Earth orbit the Sun.

Isaac Newton

Isaac Newton (1643–1727) contradicted Descartes' vortex theory, as it would cause the planets only to move in circular orbit. Newton then knew that planets moves around the Sun in an elliptical orbit.

As the vortex theory loses favor in explaining the cosmos and the movement of the planets and the moons, ether became relegated as a medium only for the propagation of light; hence, it became known as "luminiferous ether."

Christiaan Huygens

Christiaan Huygens (1629–1695) is a proponent of Robert Hooke's mechanical wave theory of light, which is similar to that of a water wave. He hypothesized that light is a wave propagating through ether.

Chapter 8: Overthrowing Einstein's Relativity Theories

Thomas Young

Thomas Young (1773–1829) had shown using the two-slit experiment that light is a wave similar to water waves whose waves can add or cancel each other.

Augustin-Jean Fresnel

Augustin-Jean Fresnel (1788-1827), a French engineer and physicist had shown by mathematical methods that polarization could only be explained if light was entirely a transverse wave.

James Clerk Maxwell

James Clerk Maxwell (1831–1879) in his electromagnetic theory had shown that light as an electromagnetic wave. (Electromagnetic wave is a transverse wave consists of electric wave and magnetic wave that are at $90°$ apart from each other.) Maxwell had shown that the speed of electromagnetism is the same as the speed of light.

Ether and Light

Ether is said to pervade the universe. It was inferred in Maxwell's equations that the speed of light is with respect to ether. This means that ether is the "preferred" or "absolute" reference frame, that is, a particle at rest in the ether frame would be in "absolute" rest while a particle in motion with respect to the ether would be in "absolute" motion.

In the idea that light travels in ether, in order for light to propagate at such enormous speed the ether had to be extremely rigid yet it should not impede the motion of light. Despite being considered a substance, no one had detected the presence of ether.

* * *

It is already an accepted fact that light can travel in a vacuum and do not need a medium to travel. The main argument involving ether and light is that light is still thought to travel in a longitudinal wave, like a water wave. If we follow this argument, then we can say that light is a disturbance in the ether. The way to debunk this notion is to decisively prove that light (as a photon particle, which is similar to the electron particle) is purely a particle and is not a wave or does not have a wave property. (The nature of light is discussed in Appendix A.) Light is practically similar to an electron travelling in space, that is, light is a spinning negative charge energy particle as it can be deflected by magnetism, gravitational field, electromagnetism, and free electrons in cathode ray tube.

Michelson and Morley Experiment

In 1879, Albert Michelson (1852-1931) read an article by Maxwell that mentioned the problems of detecting the ether. He took it as a challenge to detect the ether using an interferometer he developed and in 1881 used it to detect the Earth's motion relative to the ether. The results were inconclusive. In 1887, with an improved version, he tried again with the assistance of Edward Morley (1838-1923). They found nothing.

FitzGerald and Lorentz: Contraction Hypothesis

In 1889, G. F. FitzGerald (1851-1901), an Irish physicist suggested that in the Michelson-Morley experiment there could have been a contraction in the arm of the device that is parallel to the Earth's motion.

The same idea had also occurred to the Dutch physicist Henrik Lorentz (1853-1928) in 1892. He accounted for the null result by assuming that the contraction occurred because the electrical forces within a body were modified in the direction of motion through the ether. (Lorentz was the one who derived the transformation equations subsequently used by Albert Einstein to describe space and time.)

Theory of Special Relativity

According to Albert Einstein (1879-1936), the seed of his discovery of the theory of special relativity came as early as when he was only 16 years old when he thought of the question about what one would see if one travels with a beam of light. It took him ten more years to entertain the same problem again. Einstein thought that he should observe a beam of light as an electromagnetic field at rest but is spatially oscillating. (Note that his idea of light as an electromagnetic wave was still based on the wrong understanding of electromagnetism. I had discussed the light in Chapter 5 and in Appendix A.) Einstein thought that judging from the standpoint of an observer; everything would have to happen according to the same laws for an observer that is relative to the Earth that is at rest.[1]

(Regarding ether, although by 1904 Einstein had found out about the experiment of Michelson and Morley through the work of Lorentz, according to him this did not play a significant role in the formulation of his theory.[2])

Albert Einstein was very familiar with Galileo's work on inertia. One of Galileo's thought experiment was about being shut with a friend in a cabin below the deck of a ship that is sailing smoothly having some flies, butterflies, other small flying animals, and fishes. Galileo said that it would be observed that animals would fly the same way if the ship is standing still or that the fishes will swim any different. Throwing objects in either direction where the ship is going or that jumping up will not make any difference than in a ship that is standing still. That is, there is no difference in what a person can do in a ship that is moving smoothly (at a constant speed) and the person in a ship that is in a standstill or a person on land. (This argument was Galileo's defense of Copernicus' idea that the Earth is not at rest but is rotating and revolving around the Sun; this was in reference to the argument against his idea that the Earth is moving.)

Based from these explanations, it is said that Galileo had put forward the basic principle of relativity that states that the laws of mechanics is the same in all inertial frames. (Inertial frame is a frame of reference that is in a state of constant straight-line motion.) (One of the strangest things in science is how Galileo's understanding of inertia had been accepted without any question at all even if it was flimsy and unscientific.) For example, Newton's second law, $F=ma$, in one frame has the same form, $F'=ma'$, in another. It is said that the laws of mechanics are covariant, that is, they retain their form with respect to the Galilean transformation. (The Galilean transformation equations: $x' = x - vt$, where $t'=t$, shows that $a'=a$, which follows that $F'=F$.)

In a paper titled *On the Electrodynamics of Moving Bodies* published in June 1905, Einstein referred to an inconsistency of Maxwell's electrodynamics when applied to moving bodies using an example of a magnet and a conductor. Consider a wire moving at constant velocity across the pole of a magnet. In the magnet's frame where the magnet is at rest and the wire moves at a velocity, an observer in this frame says that the wire experience a magnetic force. In a wire's frame where the wire is at rest and the magnet is moving at a velocity, the observer in this frame would say that the wire experience only an electric force.

In his formulation of the theory of special of relativity, Einstein had to choose which one is correct: the Galilean transformation and the laws of the mechanics or Maxwell's equation. Based on the success of Maxwell's theory, Einstein decided that the Galilean transformation and the laws of the mechanics had to be modified. Einstein forwarded two postulates as the basis for his theory of special relativity:

Chapter 8: Overthrowing Einstein's Relativity Theories

1. The principle of relativity: All physical laws have the same form in all inertial frames.
2. The principle of constancy of the speed of light: The speed of light in free space is the same in all inertial frames. It does not depend on the motion of the source or the observer.

Special relativity states that the laws of physics are the same for any observer in uniform motion, that is, observers moving at a constant velocity. It is called "special" since it is restricted to the so-called inertial frame (also called inertial frame of reference). For example, take a person sitting on land and a person sitting on an airplane. One of them can pour a cup of coffee or bounce a ball, and have the same laws of physics apply. In fact, there is no way to determine which one is at rest or in motion. The person sitting on land could consider that he is at rest and the person on a plane is in motion, while the person on a plane could consider that he is at rest and a person on land is in motion as the Earth rotates. There is no experiment that can prove who is right.[3]

All inertial frames are in a state of constant rectilinear motion with respect to one another. Measurements in one inertial frame can be converted to measurements in another by a simple transformation: Galilean transformation in Newtonian physics and Lorentz transformation in special relativity.

Einstein's special relativity had brought forth ideas such as time dilation and length contraction. That is, space and time is not fixed anymore as believed by Isaac Newton. (Einstein was also influenced by Ernst Mach, who contradicted Newton's concept of the absolute time.)

In the 1905 paper on special relativity, Einstein famously dismissed the concept of ether as "superfluous."[4] Yet in a lecture in the later year of 1920, Einstein said that special relativity does not compel us to deny ether; he assumed the existence of ether but stripped it of any motion.[5] (Einstein would use this notion again in his theory of general relativity.)

Theory of General Relativity

Einstein realized that his theory of special relativity in 1905 was incomplete in two ways: It held that no physical interaction could propagate at the speed of light (supposed to have conflicted with Newton's idea of gravity as a force that acts instantly at a distance) and the other is that it is only applied to constant velocity motion. To generalize his relativity theory, Einstein tried to come up with a relativity theory that is applied to accelerated motion.

Einstein first got the solution in 1907 through a thought experiment about a free-falling observer that led him to think that the local effects of being accelerated and of being in a gravitational field are indistinguishable. For example, a person in a close chamber who felt his feet being pressed on the floor will not be able to tell whether he is in outer space being accelerated or he is at rest in a gravitational field.[6]

This led Einstein to the formulation of the equivalence principle that states that inertial mass and gravitational mass are equivalent. That is, inertial effects such as resistance to acceleration and gravitational effects such as weight are equivalent.

Einstein noted that one consequence of this equivalence is that gravity should bend a light beam. In 1911, Einstein published a paper in *Annalen der Physik* titled "On the Influence of Gravity on the Propagation of Light." He came up with a prediction that a ray of light going past the Sun would be deflected by 0.83 second of arc. This would lead later to the idea that the bending of light meant that the fabric of space through which light travelled was curved by gravity. That is, gravity arises from the curvature of space-time. Gravity is geometry.

Rotating motion is a form of acceleration, which according to principle of equivalence is also gravitation. Einstein with the help of his friend Marcel Grossman (1878-1936) found the answer to the work of Bernard Riemann (1826-1866), who

Chapter 8: Overthrowing Einstein's Relativity Theories

studied under Carl Friedrich Gauss. Gauss pioneered the geometry of curved surfaces, which is a non-Euclidian geometry. Riemann helped developed the metric tensors, which is a mathematical tool that tells us how to calculate the distances between points in a given space. Einstein needed a mathematical equation describing two complementary processes: one is how gravitational field acts on matter and telling it how to move and the other is how matter generates gravitational field in spacetime and telling matter how to curve.

The physicists John Wheeler described this interplay of how objects curve spacetime and in turn, how this curvature affects the motion of the objects was best described as: "Matter tells spacetime how to curve and curved space tells matter how to move."

As of 1912, Einstein was still struggling to develop a gravitational equation using tensors along the line developed by Riemann, Ricci, and others.

Einstein and Michele Besso (1873-1955) looked at whether rotation could be considered as a form of relative motion. (Besso was a Swiss/Italian engineer of Jewish Italian descent close-friend of Einstein in the Federal Polytechnic Institute in Zurich and then later at the patent office in Bern. Besso was credited for introducing Einstein to the works of Ernst Mach.)

Newton in his *Principia* described a bucket that is hung from a rope and rotating. As the bucket rotates, the friction between the bucket wall and the water causes the water to spin and assume a concave shape. Newton deduced from the spinning water in the bucket of an absolute and nonrotating space, that is, the bucket is spinning relative to absolute space. Newton believed that motion was only meaningful if measured with respect to another object.

Ernst Mach (1838-1916), an Austrian physicist and philosopher debunked this notion of absolute space and argued that inertia existed because the water is spinning relative to the rest of the matter in the universe. Einstein would later coin the term "Mach principle," which argued that a body's inertial mass is actually not a property of the body but rather that it is the product of the interaction between the body and the surrounding matter of the universe. Stated another way is that the local inertial frames are determined by the large distribution of matter—"the mass out there influences the inertia here." (I had explained inertia in Chapter 2 and Mach's inertia is simply nonsensical.)

Einstein later thought that it is the same whether the bucket is spinning or was motionless while the rest of the universe is spinning around it.[7]

In November 1915, Einstein finally finished his general relativity in a paper titled "The Field Equations of Gravitation."

How to Overthrow the Theories of Special Relativity and General Relativity

Overthrowing the theories of special relativity and general relativity is a formidable endeavor that requires a strategic thinking. (Formidable in the sense that many brilliant physicists swear to it that the theories are correct.) It is not about arguing the mathematics within these theories but it is about destroying their very foundations.

First, I had disputed our present understanding of inertia in Chapter 2 on *Disputation of Galileo's, Descartes', and Newton's Concept of Inertia* and in the following section, I will dispute Galileo's "relativity." Second, I will dispute special relativity's principle of relativity and principle of the constancy of the speed of light. Finally, I will dispute general relativity through Einstein's source of inspiration, the principle of equivalence principle. I will show that all the proofs and observed phenomena attributed to these theories can be explained by something else. I will show in the end that Einstein's relativity theories were just an illusion built on his misunderstandings.

Chapter 8: Overthrowing Einstein's Relativity Theories

Inertia and Uniform Motion

In Chapter 2, I had argued that the definition of inertia should only be:

A body in the presence of gravity will resist any movement.

Inertia is not about a body "persisting in a state of rest or uniform motion unless it is compelled to change that state by a force impressed on it." That is, there is no such thing as inertial reference frame or a frame that is in uniform motion. Under the presence of gravity, all bodies are in a state of inertia—even bodies at rest. If we "rate" the state of inertia, then an immobile body is zero and as the body increases in acceleration the "rate" of inertia also increases.

Disputation of Galileo's Relativity

In Chapter 2 on section *Disputation of Galileo, Descartes, and Newton's Concept of Inertia* I had disputed Galileo's principle of inertia in that inertia is the resistance of the body from any movement and not the persistence of a body from moving in a constant motion. This constant motion or constant velocity will play a big role in the idea of relativity that Galileo had not foreseen.

The purpose of Galileo in using an argument, which is more of a thought experiment, of a cabin under a ship is to prove that we will not know if we are at rest or in motion, just as we are now when we are spinning with the Earth. This argument was supposed to be his support of Copernicus' heliocentric model where Copernicus asserted that the Sun is orbited by the Earth and other planets. Galileo argued that we would not notice that the Earth is spinning, while it is at the same time orbiting around the Sun. The problem with this argument is that it is too flimsy to be use as a basis for relativity. His thought experiment will only work with a ship that is moving at slow constant speed. (Earth on the, other hand, is moving at a very fast speed of 1,670 kilometers per hour (1,037 miles per hour) at the equator.) If this is done with a jet plane that is moving at a constant speed, then things inside the plane will be pushed to the back of the plane as caused by inertia.

(In the sequence of Galileo's writing, the example of the "cabin under the ship" was told later than the "stone being drop from a mast of the ship.")

We do not usually notice that the Earth is spinning on its axis at a very fast speed but there is an experiment that can prove that the Earth and the things on its surface are spinning with the Earth at a very fast speed. Galileo mentioned of an experiment about dropping a stone from the top of the mast of the ship, which showed that the stone will not land at the foot of the mast but rather away from the direction that the ship is moving. That is, the stone will fall directly below from its original location.

> *Beside which, there is the very appropriate experiment of the stone dropped from the top of the mast when the ship is standing still, but falls as far from that point when the ship is sailing as the ship is perceived to have advanced during the time of the fall, this being several yards when the ship's course is rapid.*[8]

We can actually prove that the above experiment is true from another proven and much demonstrated experiment of a table with tablecloth cover and plates and glass on top of it (read Chapter 2 on section *States of Inertia: A Body on the Surface of a Gravitational Producing Body*). The tablecloth is the ship and the plates and glass are the stone. It does not matter if the plates and glass are dropped or sitting on top of the tablecloth that is suddenly pulled away. If they are dropped, they will fall directly below where they were. If the tablecloth is pulled very fast from under them, they will still

Chapter 8: Overthrowing Einstein's Relativity Theories

stay on their original location. (We can of course neglect the friction that will move the plates and glass a little bit.)

The experiment of a "stone being drop from a mast of the ship" practically contradicts the thought experiment of a "cabin under the ship." The thought experiment of a "cabin under the ship" was a completely wrong support for Galileo's argument that we will not know if we are at rest or in motion. That is, Galileo's "relativity" was completely wrong. Galileo had unintentionally birthed the idea of relativity that Einstein had adopted and that had dominated today's physics.

Disputation of the Special Relativity (Constant Velocity)

The two postulates of the theory of special relativity are:

1. The principle of relativity. This postulate states that all physical laws have the same form in all inertial frames.
2. The principle of the constancy of the speed of light. This postulate states that the speed of light in free space is the same in all inertial frame and it does not depend on the motion of the source or the observer.

Refuting either one of these postulates should be enough to bring down this theory. More than that, their effects or proofs will be attributed to something else or refuted.

Disputation of the Principle of Relativity

Galileo's so-called "relativity" was his argument that we will not know if we are in motion even when the Earth is spinning on its axis by giving an example of a thought experiment of being inside a ship's cabin where supposed to be one can do everything whether the ship is in a standstill or is in a uniform motion. Currently, Galileo's so-called "principle of relativity" is stated as: The laws of the mechanics are the same for all observers in an inertial frame of reference.

Einstein's principle of relativity is stated as: The laws of physics are the same for all observers in an inertial frame of reference. The example for this is a person sitting on land and a person sitting inside a plane. For the man sitting on land, he could consider himself at rest while the man sitting inside the plane is in motion. For the man sitting inside the plane, he could consider himself at rest while the man sitting on land is moving with rotation of the Earth. It is said that there is no experiment that can prove who is right. (By my definition of inertia, the man sitting on land is at rest and the man sitting inside a plane is in motion.)

In a twist to Galileo's experiment of a stone being drop from the mast of a ship, in special relativity's principle of relativity, an observer in the ship will see that the stone will land at the foot of the ship's mast but an observer on land will see that the stone will be observed to travel in a curve to land at the foot of the ship's mast (Figure 8.1).

When I started reading about this illustration of Einstein's relativity that supposedly came from Galileo, I had to search on what book it came from to check if it was exactly what Galileo had written. This present thought experiment found in many articles is a complete contradiction to what Galileo had said that the stone will land *away* from the ship's mast (see Figure 2.28 in Chapter 2). If the demonstration of Figure 2.21 where to result in Figure 8.1, then all the things on top of the table will fall to the floor (Figure 8.2). Why do the present writers argue that the stone will land *at the foot* of the ship's mast? This is because this practically demolishes Einstein's principle of relativity. The idea of relativity is wrong and Einstein's understanding of inertia was wrong.

Chapter 8: Overthrowing Einstein's Relativity Theories

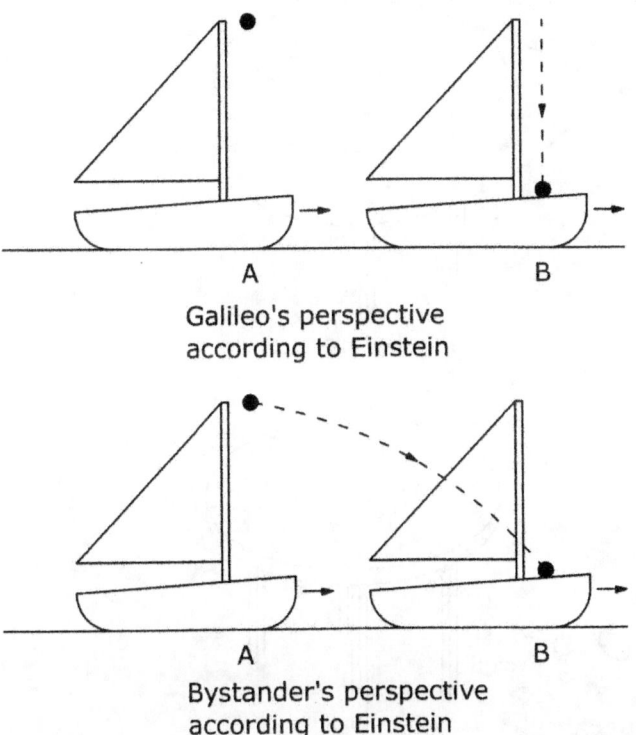

Fig. 8.1. The present interpretation of what Galileo mentioned of an experiment of dropping a stone on a ship's mast where according to Einstein's principle of relativity an observer in the ship will see the stone land at the foot of the mast while an observer on land will see the stone travel in a curve to land at the foot of the ship's mast.

THEORY OF GRAVITY

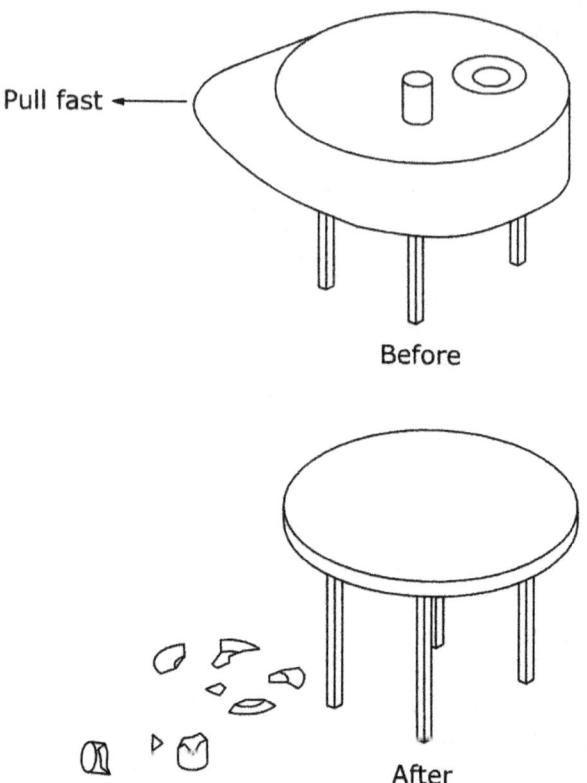

Fig. 8.2. If Figure 8.1 is applied to the demonstration of a table shown in Figure 2.21, then all the things on top of the table will go with the tablecloth. The result of Figure 8.1 shown by Figure 8.2 speaks for itself.

Disputation of the Principle of the Constancy of the Speed of Light: Ether

The speed of light had been measured to be around 3×10^8 m/s. That is, light (or rather the photon particle of light) is definitely moving. In Einstein's thought experiment of riding with a beam of light, he thought that the light he would observe would be frozen. The postulate he forwarded was that the speed of light

Chapter 8: Overthrowing Einstein's Relativity Theories

is constant and does not depend on the motion of the observer or the source (Figure 8.3).

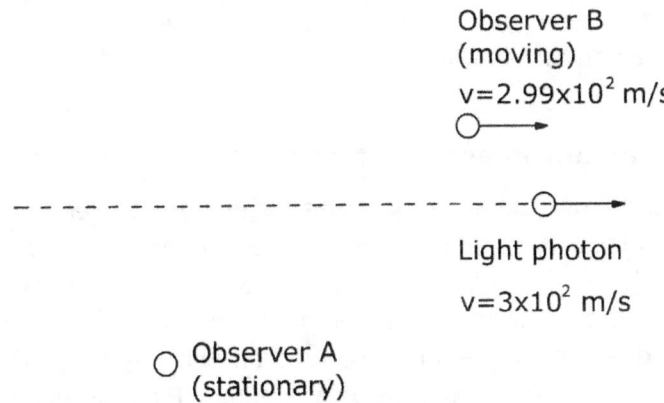

Fig. 8.3. Observer A is stationary, observer B is moving, and C is the photon particle of light. We know that Observer A will observe the particle of light at its speed while observer B will observe that light is slower.

Einstein wanted us to accept his ideas and disregard our commonsense even if our observation and mathematics (mechanics) says that it is not so. The only way that the "speed" of light could be "constant" is if light is actually frozen or not moving. (This practically contradicts the fact that light travels and therefore has a speed and that light is a particle.) This could be the reason why Einstein said that "...special relativity does not compel us to deny ether" and that "we may assume the existence of ether, only we must give up ascribing a definite state of motion to it." This is practically saying that the ether does not move. That is, the constancy of the speed of light means that light does not move. (Later in his general relativity, Einstein will transform ether into the fabric of spacetime.)

(The result of Michelson-Morley experiment did not show the existence of ether. Experiments done since that time to the present did not show any signs that ether exist.[9] We can deduce that FitzGerald and Lorentz basically believed in the existence of the ether and that their contraction hypothesis is just a means to go around the null result of the search of the ether of Michelson and Morley.)

On Electromagnetism and Relativity

As discussed in the above section *Theory of Special Relativity*, in a paper titled "On the Electrodynamics of Moving Bodies" published in June 1905 Einstein referred to an inconsistency of Maxwell's electrodynamics when applied to moving bodies. Einstein used an example of a conductor wire moving at a constant velocity across the pole of a magnet. Einstein argued that an observer in this frame would say the wire experience a magnetic force. When the wire is at rest and the magnet is moving, the observer in this frame would say that the wire experience an electric force.

First of all, Einstein's relativity is wrong. If we indulge ourselves with Einstein's arguments, his arguments have nothing to do with inertia. If we equate the magnet's magnetic field with gravitational field, his arguments affects the electron of the atom and either motion of the wire will cause an electric current (what he called as electric force) that will in turn cause an electromagnetic field (what he called as magnetic force).

Refuting the Effects of Special Relativity

There are two effects of special relativity:

1. Time dilation. The relative motion of the two frames affects the measured time interval between the two events.
2. Length contraction. The length of an object travelling at relative motion undergoes a contraction along the dimension of motion.

Chapter 8: Overthrowing Einstein's Relativity Theories

As is the case with most of the proofs of Einstein's theories of relativity, these two effects could be attributed to something else or it was the result of a wrong theory.

On Time Dilation

According to the theory of special relativity, two working clock will show a discrepancy in time when exposed under different accelerations. For example, one clock is stationary on the surface of the Earth and the other is orbiting around the Earth. It is said that it will be observed that the clock that is in orbit will lose some time compared to the clock that is stationary. It is said that the theory of special relativity was proven correct since it is used to correct the time of the GPS satellites to synchronize with the time on the surface of the Earth.

There are two things that should be noted in this observed discrepancy of the clocks: one is that the clocks use an atomic clock and the other is that one of the clocks is orbiting around the Earth. There is a very important factor in this experiment, that is, the presence of the gravitational field. (The atomic clock operates on the movement of electrons in the different atomic level wherein the electron emits a photon with certain energy associated with the wavelength and frequency or oscillations. The time is determined on how many oscillations there are to a second.)

The reason for the difference in the time between the two clocks is that the stationary atomic clock on the surface of the Earth experienced less disruption of its electrons from the gravitational fields as it is at rest with the Earth. One the other hand, the atomic clock orbiting around the Earth experienced more disruptions of the movement of its electrons from the gravitational field slowing it down as it moves with respect to Earth.

* * *

Twin paradox is a thought experiment in special relativity involving identical twins, where one travels in space and the other stays on Earth. According to special relativity, when the one that travelled in space returned, he fill find out that the one that stayed on Earth have aged more. As this is only a thought experiment, the aforementioned refutation of time dilation proves that this is not true.

On Length Contraction

According to special relativity, an object traveling at a very fast speed will have its length contract as measured by an observer travelling relative to it.

We know that when a metal heats up it expands. Plane such as an SR-71 leaks fuel prior to takeoff because it was designed to have some gap in its frame to compensate for expansion at extreme temperatures due to air friction cause by its high speed. There is no way that the length will contract when an object is travelling at a very high speed.

Overthrowing Einstein's Understanding of Gravity

In order for Einstein to supplant Newton's explanation of gravity with his theory of general relativity, Einstein had to remove the idea of gravity as a force and transform it into being mechanical and structural.

Einstein's Understanding of Gravity

Einstein argued that a free falling person inside a closed container in the presence of gravity *feels* the same as person floating inside a spaceship in space "without gravity." That is, gravity seems to be absent in a free fall. Since floating in space and falling *feels* the same, Einstein argued that there must be no force of gravity at work on both.

Chapter 8: Overthrowing Einstein's Relativity Theories

Einstein argued that a person standing in a closed container on the surface of the Earth *feels* the same as a person accelerating up in a spaceship (in an acceleration that approximates a feeling of a person standing on the surface of the Earth). That is, acceleration and gravity *feels* the same. Einstein argued that gravity and acceleration must have the same cause. (Einstein would attribute this cause to the idea that massive bodies such as the Sun or the Earth, curves space and time, and that what we call as gravity arises from this curvature.)

Overthrowing Einstein's Understanding of Gravity

Einstein's argument that a person inside a close container on the surface of the Earth feels the same as person inside a spaceship that is accelerating up in an acceleration that approximates that of what a person feels standing on the surface of the Earth. On the surface of the Earth, gravity is acting on the mass of a person experienced as weight—and this weight may not even be noticed as we grew accustomed to it. In a spaceship that is accelerating up, the amount of gravity affects the mass of a person experienced as weight as gravity pulls him down while the spaceship is going up producing an opposite force. (In all practicality, a spaceship will increase its acceleration to overcome the force of gravity so that a person is likely to be pinned down on his feet that is why astronauts are seated.) Einstein equating the feeling of a person standing on the surface of the Earth and a person accelerating up in an elevator as being the same feeling of gravity is wrong.

What Einstein was actually doing was to shift the scientific minds from their understanding of gravity as a force to the idea of gravity as caused by a massive body that creates a spacetime curvature where the other body accelerates around it. Where Einstein is going with his arguments was to frame gravity as a structure (fabric of spacetime) and mechanics (acceleration in orbit).

If we follow Einstein's arguments, then gravity is only created by a massive body and affects only those bodies that are orbiting around it while objects on the surface of the Earth are not supposed to be affected by gravity.

To say that Einstein's idea of gravity (general relativity) agrees with Newton's idea of gravity (law of universal gravitation) is wrong. To know how Einstein had derived his theory of gravity in his theory of general relativity is to know why it cannot be "merged" with quantum mechanics to derive the theory of quantum gravity since Einstein's theory of general relativity is wrong.

Disputation of General Relativity (Constant Acceleration)

When I had the idea of my theory of gravity and how to solve the problem of quantum gravity, I knew that I could overthrow general relativity on the strength that general relativity cannot solve quantum gravity or that general relativity and quantum mechanics cannot be merged to explain quantum gravity. (I wondered how the physicists have tried "marrying" general relativity with quantum mechanics through mathematics.)

Just like my theory of gravity and theory of light, both can be explained if their structure and mechanism are understood. In the case of general relativity, its structure and mechanism can be proven to be wrong. In this sense, general relativity's much vaunted mathematical proofs can be attributed to something else—proving that wrong theories can be propped up by mathematics or proofs attributed to it are wrong. (Think of the electroweak theory that was "unified" through mathematics discussed in Chapter 4.)

Chapter 8: Overthrowing Einstein's Relativity Theories

Structure of Gravity in General Relativity

The first time I started reading about general relativity, I thought that Einstein got his inspiration from René Descartes' plenum vortex (see Chapter 1 Figure 1.2). Later, I was surprised when I read that writers and physicists talks about "Newton's bucket." Newton used the water bucket analogy regarding absolute motion. What I think happened was that Einstein adopted the idea of Newton's bucket (where the spinning bucket of water made the water curve) to become the curved spacetime of general relativity as this can be inferred from his statement that it is the same whether the bucket is spinning or was motionless while the rest of the universe is spinning around it. This was later reinforced from his lecture in 1920 when he said that "special relativity does not compel us to deny ether"[10]—setting up ether as the fabric of spacetime. Furthermore, he said that, "We may assume the existence of an ether, only we must give-up ascribing a definite state of motion to it." Einstein made gravity as geometry as gravity became not a force but a spacetime curvature caused by matter.

To wit, the structure of general relativity is that of the Newton's bucket where the curved surface of the spinning water became the curvature of the fabric of spacetime made of ether but whose motion (as that of the water) had been removed. As Einstein had accorded that the rotating motion is a form of acceleration, which according to the principle of equivalence is also gravitation, his argument is set to merge with the curve spacetime as caused by a large mass.

Since quantum mechanics involves the fundamental force of gravity and general relativity is not a force at all and is wrong, it is no wonder why it had failed being unified mathematically.

(In 1905 in his special relativity, Einstein said that ether was "superfluous" and that special relativity had "blown the ether away." Yet, his general relativity in 1915 was based on the idea of an ether; and then in a conflicting lecture in 1920 he intimated that special relativity and general relativity entertained the existence of an ether.)

Mechanism of Gravity in General Relativity

Einstein's principle of equivalence states that inertial mass and gravitational mass are equivalent. This principle is where we can understand where the "relativity" is in general relativity.

Einstein's special relativity was only a "relativity" for constant velocity and so he "generalized" his theory of relativity so that it is also applied to accelerated motion. He found his relativity of accelerated motion in the case of an accelerated falling of a body and the accelerated orbiting of the Moon around the Earth or the planet around the Sun. (Here we can say that Einstein's gravity is of two parts: acceleration of a falling body and acceleration of an orbiting body. The acceleration of an orbiting body is by the mechanics of the curvature of spacetime.)

According to Einstein, a free-falling observer and an apple accelerating down obeys the principle of inertia as they are both falling together; hence, a free-falling frame of reference is an inertial frame of reference.

Also according to Einstein, a planet rotating around the Sun is actually falling towards the Sun but since it is accelerating, it follows the curvature of spacetime. Thus, gravity arises from the curvature of spacetime and is not a force. (Here we are force to believe that an orbiting body is "falling" practically infinitely.)

The "relativity" is the "inertial frame of reference" of accelerated falling is the same as the accelerated orbiting of a body. General relativity, which is the "relativity" of accelerated motion became the theory of gravity since it explained (another way) the mechanics of gravity.

* * *

Chapter 8: Overthrowing Einstein's Relativity Theories

First of all, there is no such thing as inertial mass or gravitational mass as mass is an intrinsic property of a body (matter) derived from the fundamental particles. Inertia is not gravity but it is the effect of gravity on a body. A free-falling body is under the influence of gravity as the body is being attracted by the force of gravity (polar magnetic field on the body attracted by the gravitational field), which is different from a body that is being forced into orbit by another body (field spin gravity). Another case is that a body that is moving and cutting across the gravitational field is being hindered by gravity (experiencing inertia), which is the definition of inertia. A person that is at rest will know if he is accelerating upwards just like the jetfighter pilot and astronauts in a space ship as he is pressed much harder on the back of his seat.

As Einstein had made the falling of a body and the orbit of a body as "relativity" referring to it as an inertial frame of reference, he then proceeded to changed acceleration to gravity. Notice that Einstein had used Newton's understanding of gravity in the observation of the accelerating falling apple and the accelerating orbiting Moon.

How did Einstein changed gravity as a force and changed it into mechanics? Einstein first equate the accelerating falling apple as caused by gravity but removed the cause of the falling apple as attraction due to the force of gravity (gravitational field). The falling apple or a person in a free-fall is now given the term inertial frame of reference to invoke relativity. Likewise, Einstein changed the effect of acceleration of the feeling of being stuck as caused by gravity and since he had removed the idea of gravity as a force, acceleration itself is now gravity. (Now strain your understanding and incredulity that while acceleration is a motion, the inertial frame of reference connotes as if the falling person and the apple are not moving.)

(Here we should say that acceleration or the effect of acceleration is not gravity but the effect of gravity. If a body has no means of propulsion, then it is solely through its mass that gravity can effect to accelerate it.)

Next, Einstein equated the acceleration of the Earth in orbit around the Sun as if the Earth is falling towards the Sun. Since accelerating fall was equated to gravity, then the orbiting of the Earth around the Sun, which is falling of Earth around the Sun, is gravity. Accelerated orbiting by a body, which is falling towards the body being orbited, is now gravity. Orbiting is now also an inertial frame of reference.

Einstein at last had removed gravity as a force and replaced it as a mechanics of a body, that is, of a mass. Just like Galileo, Descartes, and Newton that cannot explain why the Moon orbits around the Earth or the planets around the Sun, Einstein had used the mechanics of falling as the cause of orbit of a body. (The cause of the spin of the star, planet, and moon, which for the Sun caused the planets to orbit around it and the Moon to orbit around Earth was discussed in Chapter 3 under subsection *Theory of How Spin is Started* under section *Macro-scale/Macro-universe (Quantum Leap: Dipole Gravitational Body): Planet, Star, Solar System, and Galaxy*.) Physics had fallen down the rabbit hole. Black hole. Wormhole. Wonderland.

Refuting the Proofs of General Relativity

Phenomena have been attributed to the theory of general relativity as a proof that the theory is correct. I had shown that the theory of general relativity is wrong but it is not complete. The way to bring the theory of general relativity crashing down is to take out the support for it and to show that these phenomena can be explained by something else.

Gravitational Lensing

Gravitational lensing is said to be the proof of the theory of general relativity whereby a large body such as a star or a galaxy bends the fabric of spacetime thereby bending the path of light. This is farther from the truth.

Chapter 8: Overthrowing Einstein's Relativity Theories

In my book *Theory of Everything* that I had repeated here in Chapter 4 I had written that the photon of light and electron are practically the same; they are both spinning energy particle with a negative charge (spinning in a clockwise direction looking in front of the particle). Particles with the same charge repel each other. In the cathode ray tube experiment, the beam of cathode rays, which are free electrons, is bent (repelled) by an electric field, which is a stream of electrons. The cathode rays are also repelled by a magnet whose mediating particle of its magnetic field is the photon.

Michael Faraday's experiment with a polarized light showed that it can be bent by a magnet. Light is a photon just as the mediating particle of the magnet is also photon. Up to this point, we already know that the mediating particle of gravity is also photon. The star's and planet's magnetic field is actually their gravitational field. Therefore, we can say that the strong gravitational field (photon) of a large body such as the star or a galaxy is the one that bends or rather deflects the photon of light. (Not only that, the orientation of the gravitational field could bend the light into different directions.)

Orbit of Mercury

The gradual shifting of the orbit of Mercury over time according to general relativity is said to be due to the curvature of space-time around the massive Sun (Figure 8.4).

The reason why the orbit of the Mercury is shifting in the direction of its orbit to the Sun is that every time it passes away from its perihelion, it is being forced to shift by the gravitational field of the Sun. This can be called as orbital precession. Note that this shifting of the orbit of Mercury follows the direction of its rotation and the same as the Sun's rotation, that is, in a counterclockwise rotation. An example of this is a discus thrower who if he does not correct it, his trajectory is being shifted in the direction of his spin.

THEORY OF GRAVITY

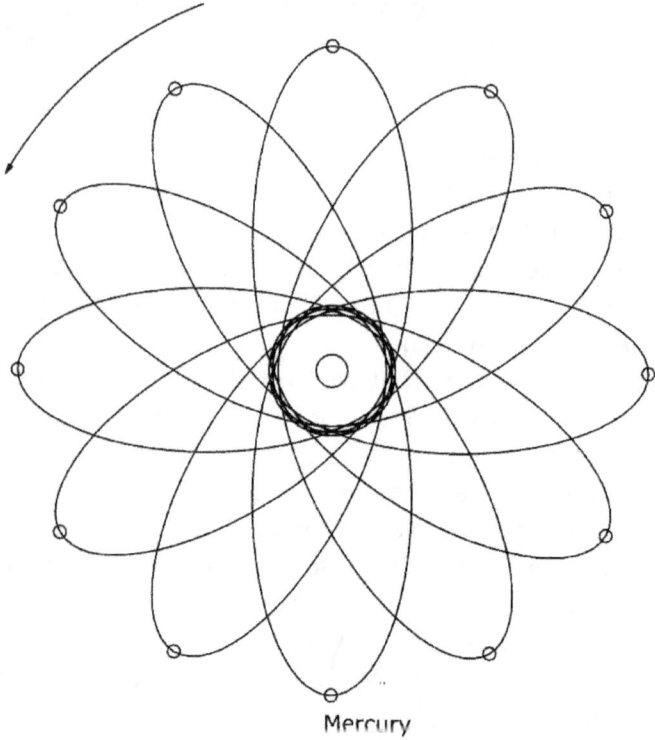

Mercury

Fig. 8.4. Mercury orbit around the Sun. Mercury's orbit is actually almost circular but exaggerated here for the purpose.

General Relativity's Spacetime Curvature and Elliptical Orbit of the Planets

Can general relativity really explain the elliptical orbit of the planets, as does Newton's law of universal gravitation, or my theory of gravity?

If the planets are riding on the spacetime curvature created by the Sun, then all of the planets angle with respect to the ecliptic of the Sun should be increasing, yet this is not the case (Table 8.1). Notice that there is no pattern in the ecliptic angle of the orbit of the planets.

Chapter 8: Overthrowing Einstein's Relativity Theories

Table 8.1 Planets: ecliptic angle and mass.

Planet	Angle	Mass (kg)
Mercury	7.01°	3.30×10^{23}
Venus	3.39°	4.87×10^{24}
Earth	0°	5.97×10^{24}
Mars	1.85°	6.42×10^{23}
Jupiter	1.31°	1.90×10^{27}
Saturn	2.49°	5.68×10^{26}
Uranus	0.77°	8.68×10^{25}
Neptune	1.77°	10.24×10^{25}
Sun	-	1.99×10^{30}

In fact, there is no current explanation as to the ecliptic angle of orbits of the planets to the Sun (Figure 8.5).

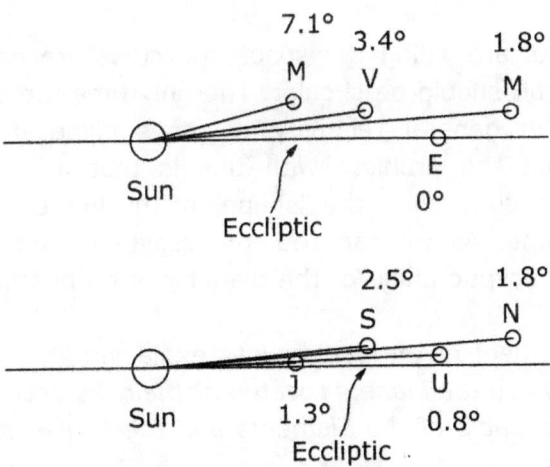

Fig 8.5. Comparison of the ecliptic angle of orbits of the planets to the Sun and the supposed general relativity's ecliptic orbit of the planets based on the spacetime curvature created by the Sun. (M=Mercury, V=Venus, E=Earth...)

Another argument to general relativity's spacetime curvature is that the heavier bodies should have been closer to the Sun, yet there is no explanation (until now) on the distribution of the mass of the planets with respect to its distance from the Sun (Figure 8.6).

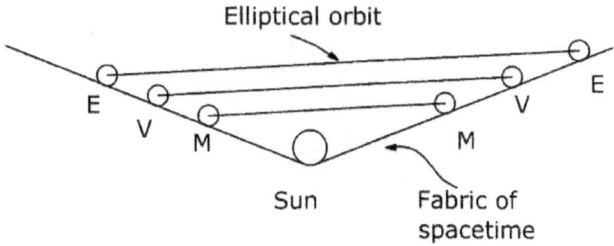

Fig. 8.6. Comparison of the mass of the planets to general relativity's spacetime curvature created by the Sun.

If the planets are riding the spacetime curvature around the Sun, their orbit should be circular. The only time the orbits of the planets in general relativity becomes elliptical is if Mercury rises but the problem with this is that it will fall opposite it, which contradicts the "shape" of the funnel of the fabric of spacetime. As we can see, the reality is Table 8.1 showed that the ecliptic angle of the planet has no bearing on its mass.

(The arrangement of the planets was explained in Chapter 3 on section *Why are the Planets Located at their Distance from the Sun: Periodic Table of the Elements and the Gravi-tational Field*.)

Black Hole

Black hole is said to be a phenomenon as a consequence of the theory of general of relativity whereby a massive body creates a deep spacetime curvature often illustrated like a funnel. It is sometimes claimed that there is a black hole at the center of

Chapter 8: Overthrowing Einstein's Relativity Theories

galaxies, particularly those of the "flat" galaxies like elliptical galaxy and spiral galaxy. A black hole is often erroneously depicted as a physical hole.

If there is black hole at the center of a galaxy, the spiral or elliptical galaxies should not be flat with a bulge in the center rather but should be shaped like a funnel. Black hole and the universe should look weird with general relativity.

(It is also erroneously propagated that CERN's LHC particle accelerator could create a black hole from the collision of particles, which is utterly nonsense since a black hole is supposedly only created by a massive body. To indulge in the idea of CERN's LHC creating even a minuscule black hole is ridiculous and not even should be entertained for publicity.)

The reality is that there is no black hole. To explain how a massive body occupies a space in the universe, a star, a galaxy, or collapsed star emits a very strong gravitational field whose structure is like that of a dipole magnet (that spins) where one pole emits the gravitational field and the other pole takes in the magnetic field. If for anything else, a massive body could grab bodies that come close to it and its very strong gravitational field could crush that body and breaks it down by its gravitational field (attraction) and by its heat. (This could be true with the formation of a galaxy from a nebula.)

Gravitational Waves

As proposed by Albert Einstein in 1915 based on his theory of general relativity, gravitational waves are supposedly produced by violent events such as the collision of black holes. Gravitational waves are said to be the ripples in the curvature of spacetime that propagates as waves, travelling outwards from the source.

In September 14, 2015, the Laser Interferometer Gravitational Wave Observatory (LIGO) captured the signal of the gravitational waves. The discovery was announced on February

11, 2016. In May 2, 2016, a Special Breakthrough Prize in Fundamental Physics was awarded to the scientists and engineers who contributed to the detection of the gravitational waves.

* * *

At this point, we know now that the theory of general relativity is wrong. There is no such thing as the fabric of spacetime or its curvature and there is no black hole. So what did the scientists had detected?

Currently, it is still understood that light and transmission from the antenna are electromagnetic waves. In Appendix A, I had discussed that there is no electromagnetic waves as light and antenna transmission is mediated by the photon particle. If the magnetic field from the magnet is not radiated from its source, then we can say that it is also the same for the gravitational field. So how is it possible for the photons of the gravitational field to be radiated out in space? Based from the description of the gravitational waves, gravitational waves can be produced by the collision of black holes. We know that black hole do not exist and that physicists described the center of the galaxy as having a black hole. We know from the definition of a black hole that it is caused by a massive object and that we can infer that at the center of the galaxy is a very massive star. Based from this, the collision of two galaxies could only mean the collision of two massive stars, which could only result into an explosion. An explosion means the release of photons out in the space, which could result into what the scientists had detected and thought to be the signal of the gravitational waves.

Chapter 8: Overthrowing Einstein's Relativity Theories

Gravitational Redshift

Gravitational redshift happens when the electromagnetic radiation originating from the source is stretched or reduced in frequency.

* * *

Note that the term electromagnetic radiation is that of the current understanding of the electromagnetic radiation spectrum from the gamma rays to the radio waves. At this stage, we already know from my theory of everything in Chapter 4 that electromagnetism, magnetism, light, strong interaction, and gravity is mediated by photon. If ever this happened, the energy of the photon could be in the gamma rays and was reduced to a lower energy range of the electromagnetic radiation spectrum. What this means is that if the source is the Sun that emits an electromagnetic radiation (photon) in the energy of the gamma rays (high frequency/short wavelength) and as this photon passes through the Sun's gravitational field, the photon collides with the photons of the gravitational field reducing the energy of the photon (low frequency/long wavelength), thus the photon could be described as being "stretched" (from short wavelength to long wavelength). Think of shooting an arrow from underwater or shooting an arrow in the water where the speed (potential energy) of the arrow is reduced.

Commentary

Among the most formidable theories to overthrow are Albert Einstein's theory of special relativity and theory of general relativity since they were supported (supposedly) by mathematics and proofs (observations) and that a great many brilliant physicists swear by it and the genius of Einstein.[3.5]

Chapter 9

Overthrowing Higgs Boson, String Theory, and the Question on the Dominance of Matter over Antimatter in the Universe

Higgs Mechanism, Higgs Field, and Higgs Boson: Ether-Based Theory

The first time I read about the theory behind Higgs mechanism that supposedly gave mass to all particles I thought it was beyond my understanding. I watched some videos where the physicists were asked to explain it and they seemed to stumble, stammer, tried so hard to explain it clearly, unable to explain it clearly, or answered it in a contrived manner. The theory of the Higgs mechanism had gone from a theory that was at first not taken seriously when it was first proposed to a theory that launched the biggest wild goose-chase with the building of billions worth of a machine who was and should be destined not to fail. How was the Higgs mechanism explained?

Higgs Field as Explained in Physics

The usual explanation for what precipitated the search for the Higgs boson was that the standard model does not take into account how the particles got their masses. It was actually due to the gauge theory, which is a branch of the standard model concerning the force carrier particles that says that all force carrier particles should be massless. The idea that force carrier particles should be massless came in conflict when it was found that the force carrier particles of the weak interaction, the $W^{\pm}$ and Z^0 have masses.

In 1957, Julian Schwinger (1918-1994) lectured and wrote about an idea of constructing a theory where the weak interaction and electromagnetism could be unified. In 1961, one of his students, Sheldon Glashow (born 1932) took up the challenge and came up with a theory but added a term that gave mass to the weak field. Glashow's theory did not have the gauge symmetry and have problems with infinities that could not be renormalized by any known methods.[1] To fix the problems of gauge symmetry the idea of the Higgs field was needed.

The theory of Higgs mechanism, which posits of the Higgs field, was based on the mathematical work developed and proposed by Peter Higgs, Robert Brout, Francois Englert, Gerald Gularnik, C.R. Hagen, and Tom Kibble. (Peter Higgs (born 1929), Robert Brout (1928-2011), Francois Englert (born 1932), Gerald Gularnik (1936-2014), C.R. Hagen (born 1937), and Tom Kibble (1932-2016).) What is not often explained in most articles was: Where did the Higgs field came from? According to one explanation, when the universe was hot the Higgs field exerted a wild influence but once it cooled down enough to below a certain temperature, the Higgs field assumed a certain value that corresponds to the lowest energy level and this energy level continues to prevail throughout the whole universe.[2]

In 1967, Steven Weinberg (born 1933) used Glashow's model to unify the weak interaction and electromagnetism but added the idea of the Higgs field so that the quanta of the weak interaction would be massive while maintaining the gauge symmetry of the theory. Abdus Salam (1926-1996) also came up independently with the same idea. The unification of the weak interaction and magnetism became known as the electroweak theory.

In 1979, Glashow, Weinberg, and Salam were awarded the Nobel Prize in Physics for the unification of the weak interaction and electromagnetism. In 1999, Gerardus 't Hooft (born 1946) and Martinus Veltman (born 1931) were awarded the Nobel Prize in Physics for showing that the electroweak theory was renormalizable.

Higgs Field as Explained to a Layperson

In 1993, UK Science Minister William Waldegrave offered a prize, a bottle of champagne, to the one who could come up with the best explanation of the Higgs field and Higgs boson to a lay person. Physicist David Miller of the University College of London won the bottle of champagne by using this analogy: In a cocktail party full of physicists (the analogy of the Higgs field), when a popular person (a particle) wanders through the crowd, he will be mob by the crowd making his passage more difficult (hence he will have a higher mass); while if a less popular person enters the room, only a small crowd gathers making it easier for him to move (hence he will have a lower mass).

Higgs Mechanism as Depicted

Later, I found more information about the Higgs field from an article in *Times* magazine issue of March 26, 2012, with the depiction of the Higgs field (Figure 9.1).[3]

(In this depiction of the Higgs field, I recognized what the Higgs field was—an ether. Note that the Higgs field is supposedly everywhere, supposedly so is the ether.)

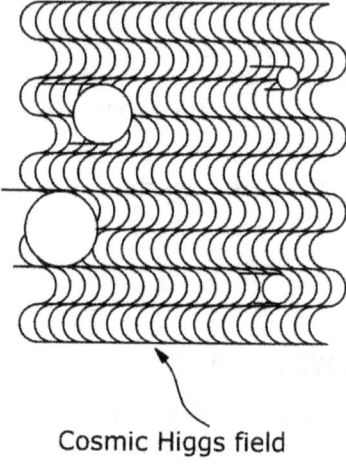

Cosmic Higgs field

Figure 9.1. A depiction of the Higgs field.

Higgs Discovery Announced and Higgs Awarded the Nobel Prize in Physics

Just over four months after reading the article in *Times* magazine, as if as a friendly ribbing on the now decommissioned competitor, the Tevatron of Fermilab in Batavia, Illinois, CERN found the Higgs boson and announced it on July 4, 2012. However, they were not totally sure yet. It was not until March 14, 2013 when CERN finally confirmed that what they indeed had found was the Higgs boson.

With the speed by which the discovery of the Higgs boson was accorded the Nobel Prize on October 8, 2013, it may be said that it was nominated long before it was announced that the Higgs boson was found. There were some who questioned the veracity of the experiment that produced the evidence of the Higgs boson. However, any objections are no match when it is defended heavily and crowned with awards. CERN's LHC was riding a tiger.

Higgs Field Explanation as the Source of Mass of the Particles

In most articles in physics, it says that the Higgs field is the source of mass of **all** the particles. It is thus mind-boggling to know that this is not all true as it was pointed out by the director of CERN that the Higgs boson proposed by Brout, Englert, and Higgs accounts **only** for the masses of the fundamental particles but that the bulk of the masses come from the binding energy of the strong interaction holding them together.[4] (You do not hear or read this information quite often. As discussed in Chapter 2 on section *Bond Angle and Gravity* shown by Figure 2.10 and subsection *Polar Gravity (Gravity-y): Repulsive and Attractive Force* under section *General Theory of Gravity* shown by Figure 2.12, strong interaction as it relates to the number of proton in an atom explains more for the weight of a body and not its mass.)

The Reality of Operating the LHC Regarding Jobs

To understand the predicament that CERN is facing, there are more than 10,000 scientists working directly or indirectly on the four experiments in LHC and operates on a budget of nearly $1B.

(On the other hand, had the Superconducting Super Collider (SSC) was built, it could have also incurred a similar amount to operate it at the possible detriment to other research and possible closure anyway at it could have been inevitably affected by economy.)

Limitations on the Use of the Particle Accelerator

The history of particle physics follows the saga of the discovery of the subatomic particles (electron, proton, and neutron). What follows next is the discovery that inside the proton and neutron are still smaller particles called quarks (up quark and down quark). This gives us the fundamental particles of up quark, down quark, and electron, and together with the determination of the fundamental forces of nature constitutes the theory of the standard model.

While the particle accelerator is instrumental in determining the basic constituents of the particle of matter, it is now also being used to "shed light" on the moments after the Big Bang, the existence of graviton, dark matter, super-symmetry, and a slew of other ideas or theories. Commonsense tells us that there must be a sensible limitation on the use of the particle accelerator.

Since the Higgs field is thought of as a space-filling field, the physicists thought of colliding particles using the LHC particle accelerator to "jiggle" the Higgs field to emit the Higgs boson, which could be recognized by the particles it decays into. For the CMS it was observed that the Higgs boson decayed into two photons while in the ATLAS it was observed to have decayed into four muons (said to be the third generation of electron of the family of lepton). There are a lot of questions to these results. Who decides that the different results (photons and muons) are the decays of the Higgs boson? If it is a Higgs boson then it may be understandable if it decays into photons but to say that it decays into electron is like saying that the proton or neutron decays into an electron or a photon. If the Higgs field is thought of as a space-filling field then there is one like it that was also thought of as a medium pervading the whole universe that is called the ether. Why then not find it like the ether.

Refuting the Existence of the Higgs Boson

From the section above on *Higgs Field as Explained in Physics*, we can refute the existence of the Higgs boson by my claim that the Big Bang did not actually happened and that in my discussion in Chapter 4 under subsection *On Electroweak*

Chapter 9: Overthrowing Higgs Boson, String Theory...

Theory under section *Fundamental Forces* that the weak interaction and (the old understanding of) electromagnetism cannot be unified as they are both distinct from each other unlike strong interaction, electromagnetism, magnetism, gravity, and light.

In the idea of the fundamental forces of nature, if the Higgs boson is the mediating particle of the Higgs field, which gives mass to the other particles, then mass should have been also a fundamental force of nature. I had stated in my book *Theory of Everything* and in this chapter that mass is an inherent property of an energy particle derived from its motion ($m=E/c^2$). We know from Chapter 4 that all of the fundamental forces of nature have its source from the fundamental particles. Quark strong force and quark weak force (strong interaction), atom strong force and atom weak force (residual strong interaction or strong nuclear force), magnetism, electromagnetism, and gravity came from the energy field of the fundamental particles. (Photon of light and neutrino are emission from the fundamental particles.) Fields can only be created by a dipole magnet energy particle (quarks, proton, neutron, and electron in an atom); electrons in a wire; or by the dipole magnet configuration of the planet, star, and galaxy in the form of gravitational field. That is, there is no field that exists outside of those created by the fundamental particles. The most important question on the existence of the Higgs boson is:

If the Higgs boson of the Higgs field exist, then where and what particle is the source of the Higgs field?

Where did the Higgs field derived its field from and where is this source particle, which can only be a dipole magnet energy particle? If the LHC had found the Higgs boson then where did it originated? (At this point, count out the dark matter and dark energy, as they are both the mechanism of gravity.) Note that when protons are smashed against each other, the fields of the quarks inside the proton are, for all intents and purposes, also made up of the photon. That is, those produced by the LHC particle accelerator could easily be the photon from the energy field of the proton.

Which one is a much simpler to explain as the source of mass of the particle? The one where it was given to them by another particle (Higgs mechanism) or the one that is their inherent property derived from their motion (my Mass theory) as given by the established Einstein's equation, $E=mc^2$, which shows that mass, m ($m=E/c2$). That is, mass is the property of a particle based from its amount of energy and its speed or motion. Now compare the Higgs analogy against Einstein's formula on mass, $m=E/c^2$:

1. A particle with an energy E that is moving slow will have a lighter mass while a particle with the same energy E that is moving fast will have a heavier mass m.
2. A particle having more energy E' and is moving at the same speed c will have greater mass m' compared to a particle having a lesser energy E'' that will have a lesser mass m''.

The reality is that the effects of bringing down the Higgs boson would be to free up large resources (money, brain, and time) that could be used to solve other problems.

String Theory: Graviton

The string theory had been studied in the 1960s as a theory of strong nuclear force before it was abandoned in favor of the theory of quantum chromodynamics (QCD). It was then realized that the very properties it was made unsuitable for nuclear physics made it suitable as a candidate for the theory of quantum gravity.

Chapter 9: Overthrowing Higgs Boson, String Theory...

In spite of the failure of the string theory in nuclear physics, it wormed its way into the quantum mechanics in the standard model by positing to replace the point-like particles of particle physics (standard model) with a one-dimensional object called string.

According to string theory, a string oscillates in many ways, which gives rise to a different species of particle with its mass, charge, and other properties determined by the string's dynamics. One of the modes of oscillation of the string corresponds to a massless spin-2 particle of which the graviton was theorized to have the same property. In the standard model, graviton is said to be the mediating particle of gravity. Since the string theory is said to be consistent with quantum mechanics, the discovery of the graviton implies that the string theory is a theory of quantum gravity. Some of the claims of the string theory are:

1. It claims to incorporate gravity; hence, it is a candidate for the "theory of everything."
2. It is now widely used as a theoretical tool and had shed light on many aspects of quantum field theory and quantum gravity.
3. It accommodates a consistent combination of quantum field theory and general relativity.
4. It agrees with the insight on quantum gravity such as the holographic principle and black hole thermodynamics.

Notable critics of the string theory are: Richard Feynman, Roger Penrose, Sheldon Lee Glashow, Peter Woit, Lee Smolin, Philip Warren Anderson, Lawrence Krauss, Carlo Rovelli, and Bert Schroer. Some of the common criticisms are:

1. It needs a very high energy to test quantum gravity.
2. It is lacking unique predictions due to large number of solutions.

3. The lack of background independence.

(Background independence is a condition in theoretical physics that requires the defining equations of the theory to be independent of the actual spacetime and the value of the actual fields in the spacetime, that is, the theory must be coordinate-free.)

In layman's term, the theory cannot be tested nor can the theory predict anything.

As I had discussed in this book, the following are the reasons why the string theory fails:

1. The string theory cannot even explain gravity. As it is claiming to be a candidate for the theory of quantum gravity through the still to be discovered graviton, it failed since graviton does not exist. Thus, string theory also failed as a candidate for the theory of everything.
2. As a theory that utilizes general relativity, it also failed since Einstein's general relativity is wrong.
3. Its claim of "insight" into the black holes is wrong since general relativity is wrong.

Currently, string theory is also into the search for super-symmetry (SUSY), dark matter, dark energy, and the idea of a rapidly expanding universe. String theory is a theory grasping for straws.

The Alliance of the Higgs Boson Theory and String Theory Theorists: Their Strategy and the Use of the Particle Accelerator

I never thought before that the string theory is only connected to the standard model through the theory of the existence of graviton, which is the theorized mediating particle of gravity. Upon watching Sean Carroll's DVD *Higgs Boson and Beyond*, I was surprised to learn that the supporters of the string theory are also the driving force in the pursuit to discover the Higgs boson. The string theorists found their foothold in the weakness of quantum mechanics and in the quantum field theory. In quantum mechanics, it is still understood that particles have a wave property. (In Appendix A, I had shown that the photon of light is a particle with no wave property whatsoever just like all the fundamental particles, the subatomic particles, or their mediating particles.) In the quantum field theory, it is thought that what we perceive as particles are thought to be the vibrations in the fields that pervade all of space. That is, the string theory posited that these particles vibrate, which suits beautifully in their theory of the particles as a string that vibrates.

The Question of the Dominance of Matter over Antimatter in the Universe: Big Bang Theory

In physics, the idea of the dominance of matter over antimatter in the universe is called matter-antimatter asymmetry. The question of why matter became dominant over antimatter can only be taken into account in the light of the Big Bang theory. In this book, I had overthrown the Big Bang theory with my Hydrogen Origin Theory (HOT) of the Universe.

To explain the dominance of matter over antimatter in the universe is to explain that without thinking of matter and antimatter our universe was created with the established spin/charge of the fundamental particles: up quark, down quark, and electron. That is, our universe was built the way it is. The existence of the antimatter is even subject into question as there is no known existence of the opposite spin/charge of the up quark as the nucleus and down quark as the one orbiting the up quark in the proton and neutron, and the opposite spin of electron. That is, there is no antimatter. The only so-called "antimatter" particles are free particles, which were observed to spin the opposite of their normal spin usually due to collision, nuclear fission, or nuclear fusion. It is usually through collision that a particle could be able to spin opposite to its original spin, which sooner or later it is either to return to its original spin or will hit another particle that will result into a massive release of energy. The latter is caused more by the high potential energy imparted on that particle. For antimatter to be real, it has to be an atom where the fundamental particles spin (their charge) are in the opposite direction.

Epilogue

The fields of particle physics, gravitational physics, astrophysics, and cosmology are lost right now as can be observed by the published articles in print and in the internet where there is no sense of exploration anymore but desperation to find answers of the problems in physics by speculations. The solutions I had offered here should bring these fields of physics back to the right track and back to the reality. For my theories and ideas to flourish, it is not enough to put forth theories and ideas out there in the world if nobody would look at it, talk about it, and when proven right to champion it.

It remains to be seen how the world will react to this change—to stay in the darkness or to herald the coming of the new age.

Appendix A
Theory of Light

This appendix is a new development from my unpublished book on my theory of light that will completely overthrow of the wave property of light or the dual wave-particle property of light to establish light only as a particle, a photon energy particle. Overthrowing the wave property of light and all particles is very important in removing the wave property of particles in quantum mechanics. This is also the fulfilment on my revised standard model and my theory of everything.

The Nature of Light: Past to Present

The nature of light had been argued even in the times of the Ancient Greeks. Light is thought of as a particle, a fire in the four elements of nature (fire, earth, air, and water), it travels in a straight line, and its source is in the eye.

During the Renaissance period (14th to 17th century) brings the revival of the knowledge of the ancient Greeks and Romans (and more possibly with the influence of the ancient Indians), which was bridged by the Islamic civilization. Isaac Newton (1643-1727) proposed a corpuscular theory of light in which light is thought to be a stream of particles he called "corpuscles." Christiaan Huygens (1629-1695) proposed that light is emitted in all directions as a series of waves in a medium called ether. Huygens' wave propagation of light became the basis of the wave theory of light. Thomas Young (1773-1829) proved the wave theory of light by demonstrating the wave nature of light by simulating the interference produced by two sources of water waves in a historic experiment called double-slit experiment. (Newton's particle theory of light held sway up to the time of Young's experiment.)

It was Albert Einstein (1879-1955) in his explanation of photoelectric effect that resurrected again the particle theory of light. Einstein asserted that light has both a wave and particle property (wave-particle duality) and that one can bring out one or the other depending on the experiment done. (Young's double-slit experiment shows the wave property of light and Einstein's explanation of photoelectric effect shows the particle property of light.)

Light as an Electromagnetic Wave of an Electromagnetic Radiation Spectrum

Light is considered as an electromagnetic wave within the range of the electromagnetic radiation spectrum (Figure A.1). The light that we often referred to is called visible light.

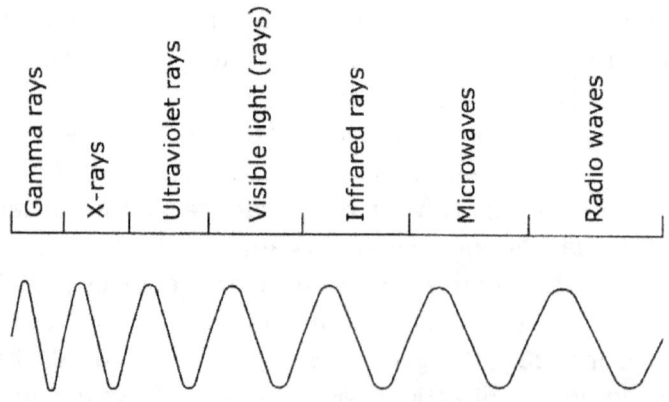

Fig. A.1. The electromagnetic radiation spectrum.

Appendix A: Theory of Light

My Previous Understanding of the Structure of Light From My Unpublished Book on My Theory of Light

In my unpublished book on my theory of light I had discussed how I had the insight to plot the electromagnetic radiation spectrum (gamma rays, x-rays, ultraviolet rays, visible light, infrared rays, microwaves, and radio waves) by dividing them into three parts: from gamma rays to ultraviolet rays, the visible light, and from infrared rays to radio waves. I plotted the frequency in the y-axis (the direction that the particle is moving) and the wavelength in the x-axis (the energy field radiated by the particle). I accorded the spin on the particle to give it a charge (charge and spin are the same) and to somehow explain the wave property of the particle (Figure A.2).

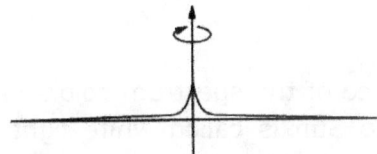

Fig. A.2. My previous understanding of the structure of light from my unpublished book on my theory of light.

(If my memory serves me, I derived the direction of the spin of the photon of light from the direction of the spin of the electromagnetic field. Note that even until now, the electron is not thought to spin inside the wire conductor.)

Think of the wake of the ship in the water or the shockwave of the bullet or a jetfighter. It took me to finish my book *Theory of Everything* and this book to discard radically this theory on the structure of light. As discussed in Chapter 4 under subsection *Light* on section *Fundamental Forces*, the photon particle of light traveling in space does not emit an energy field. (It is also known that the photon of light inside the fiber optic does not emit an "electromagnetic field.")

Theory of Light

Until now the nature of light is still not known. Light is thought to be a particle, a wave, or having a dual wave and particle property. There are many misconceptions on light:

1. Light is said to contain all the colors of the visible light as shown by the spectrum from the prism.
2. When a light hits an object, the color of the object that we see is the color that bounces off the object while the rest of the colors are absorbed.
3. Light has a frequency and wavelength.

Correcting the misconceptions on light answers the nature of light.

Prism

The light that is often the source of the spectrum colors on the prism that is coming from the Sun is called white light. The light that is coming from the Sun after passing through our atmosphere is often reduced to the range of the visible light or above the energy of the visible light, that is, in the energy range of the ultraviolet rays. When the sunlight passes through a prism, the prism does not separate the colors of the white light from violet to red, as previously thought. That is, the light does not contain the colors of the spectrum. Rather, when the sunlight passes through the prism, the photons of the sunlight hits the electrons of the atoms of the prism, where the electrons vibrates emitting the different energy photons. (Some of the photons of light may even pass through the prism and retain their high energy seen as white light.) That is, we perceived the energy of the photon as a color. (In my unpublished book on my theory of light, the prism actually acts as a slit where the light source is projected at both ends of the slit making the spectrum looks oval.)

Appendix A: Theory of Light

Note that the higher the energy of the photon, the greater is its mass ($E=mc^2$). That is, blue has a higher mass than red. In what I called energy-mass dispersion, the higher is the energy of the photon the shorter it is thrown. This is the same as throwing a rock, where a heavy rock can only be thrown at a short distance. Taking into account the direction of the source of light and the orientation of the prism, blue is thrown closer to the prism and red would be further.

Color

Color could either come from the source or the light from the source hits a body to show the color of the body. Coming from a source, we can actually ascertain the temperature of the source based from its color: blue is hot and red is less hot. The most misunderstood is explanation of the color of an object.

Currently, it is understood that when the light hits a body, the body absorbs all the colors of the light and bounce off the color that we perceive as the color of the body. This explanation is actually wrong. When the light hits the body, the photon of the light hits the electrons of the atom of the body. Some of the energy of the photon of light is absorbed by the atom, which is experienced as the heat absorbed by the body. Some of the photon of light that hits the electrons of the atom dislodges the electron from its orbit and upon its return to its orbit, it releases a photon corresponding to the color we perceived. Thus, the color of the body we perceived is actually the energy of the photon emitted by the electrons of the atom. Hence, the color of the body has to do also with the property (atom) of the body.

Frequency and Wavelength

From the above subsection *Color*, the color of the body is actually the energy of the photon emitted by the body. The frequency we associate the color with or the whole range of the electromagnetic radiation spectrum is explained by the formula, $E=hf$, where E is the energy of the particle, h is Planck constant, and f is the frequency. That is, the frequency was conventionally associated with the energy of the particle. The wavelength we associate the frequency of the light is given by the formula, $\lambda = c/f$, where λ is the wavelength, c is the speed of light, and f is the frequency. Since light is an energy particle, there is actually no frequency or wavelength. (The issue of frequency and wavelength is much better understood in the subject of antenna discussed below.)

Antenna

It is currently an accepted and established knowledge that an electromagnetic radiation particularly used for electronic communication is propagated through electromagnetic waves. That is, electromagnetism is thought of as waves, particularly transverse waves of electric field and magnetic field. In the same way that light is a photon and that the mediating particle of electromagnetism is photon then the photon of electromagnetism is purely a particle and has no wave property.

Antenna and Electronic Communication: Sending and Receiving Information

An electromagnetic radiation is propagated by an antenna. Long distance electronic communication commonly uses the range from microwaves to the radio waves. The information is contained in the energy of the electron that flows in the wire and is sent to the antenna where in the antenna it creates an electromagnetic field that is emitted (transmitted) to space. Think of the series of photons of the electromagnetic field emitted as like a conveyor belt containing the series of information as the energy of the photon.

Appendix A: Theory of Light

When the photon hits the receiving antenna, the photon hits the electron of the antenna where it will make the electron flow. The information is contained in the series of photons emitted by the transmitter antenna that hits the receiver antenna that is then replicated in the series of electrons received by the receiver.

Where is the Frequency and Wavelength

The frequency of an energy particle had been derived from the formula: $E=hf$, where E is the energy of the particle, h is the Planck constant, and f is the frequency. Consequently, the wavelength is derived from the formula: $\lambda=v/f$, where λ is the wavelength, v is the speed and f is the frequency.

What happened is that for the electromagnetic radiation spectrum, the frequency had been *conventionally* derived from the formula $E=hf$. The ranges of the electromagnetic radiation spectrum can be explained without the need for the frequency and wavelength.

For the electromagnetic radiations spectrum, the "high frequency" such as the gamma ray is a high-energy photon particle while the "low frequency" such as the radio waves is a low-energy photon particle. The "high frequency" has a "short wavelength" while the "low frequency" has a "long wavelength." For the electronic communication, high energy particles ("high frequency" and "short wavelength") only needs a shorter (small) receiver antenna since the high energy photon particle could easily make the electrons flow to receive the information. Low energy particles ("low frequency" and "long wavelength") need a long (big) receiver antenna since the photon particle has a low energy.

Appendix B
Flying Saucer

We are bombarded with news or documentaries of flying saucers that are flying very fast and turn even at a right angle where no known terrestrial aircraft can do and which its occupants are supposed to experience very large order of gravitational acceleration (g) that could be lethal for humans. We do not recognize that this flies in the face of Einstein's theory of special relativity regarding inertia and theory of general relativity regarding gravity.

Our understanding of gravity will lead us to the dawn of traveling on air and in space using gravity. Our ability to build flying saucers is very important also in looking for another habitable planet for the survival of the human species.

Flying Saucer

A gravity producing body has a structure of that similar to the dipole magnet energy particle (Figure B.1).

Since there is actually no anti-gravity, the reasonable explanation of the propulsion of a gravitational vehicle is like that of the repulsion of two magnets having the same poles (Figure B.2).

THEORY OF GRAVITY

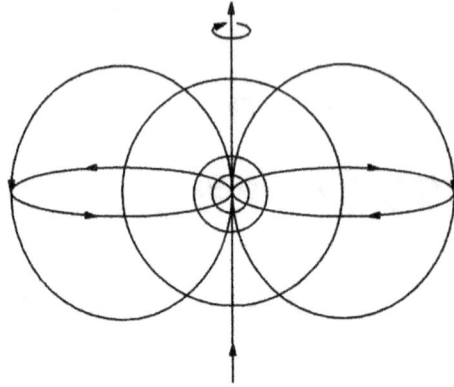

Fig. B.1. The structure of a gravitational producing body showing the source of gravity and its gravitational field. (The top spin shown is the direction of the spin of the photon of gravitational field and not the spin of the gravitational producing body.)

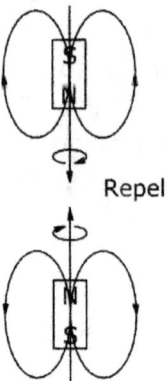

Fig. B.2. Repulsion of the same poles magnet. (The spin shown is the direction of spin of the photon of the magnetic field.)

It is interesting to note that I had always assumed that since the photon of light spins (in a clockwise direction/negative charge) that the photon of the magnetic field also spins. Also,

Appendix B: Flying Saucer

that the magnetic field of the magnet as a whole should also be spinning in the counterclockwise direction looking down from its north pole as the magnetic field of the magnet is the result of the of the spin of the up quark passed down to the proton and to the atom.

A gravitational vehicle opposes the gravitational field of the Earth (Figure B.3). A gravitational vehicle should be maneuverable, unlike the Moon that is governed by the Earth (or the planets that is governed by the Sun) where the Earth controls the Moon in a constant manner.

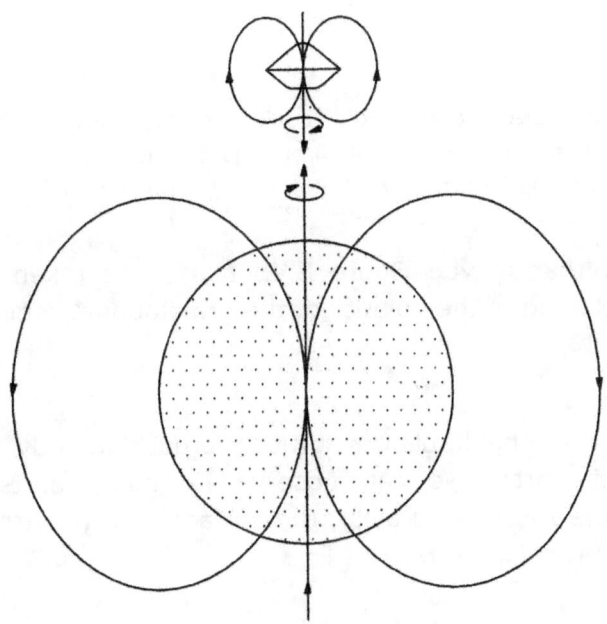

Fig. B.3. A gravitational vehicle repels the gravitational field of the Earth. (The spin shown is the spin of the photon of the gravitational field of a gravitational producing body.)

* * *

We can think of Earth's gravitational field as moving out in all directions (Figure B.4a) or based on the Earth's poles, the

THEORY OF GRAVITY

gravitational field or magnetic field moves out on the (magnetic) North Pole and enters on the (magnetic) South Pole (Figure B.4b).

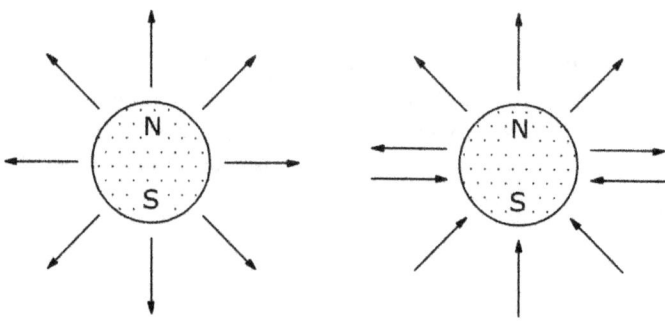

Fig. B.4. Earth's gravitational field could be viewed with: a) a gravitational field moving out in all directions b) moving out in the North Pole and moving in in the South Pole.

I thought that maybe Figure B.4b has to be taken into account in switching of the polarity of the gravitational vehicle.

* * *

Looking closely at a habitable gravitational producing body such as our planet Earth, we can observe its many parts as consisting of the source of gravity, the surface, and the atmosphere/living space (Figure B.5).

Appendix B: Flying Saucer

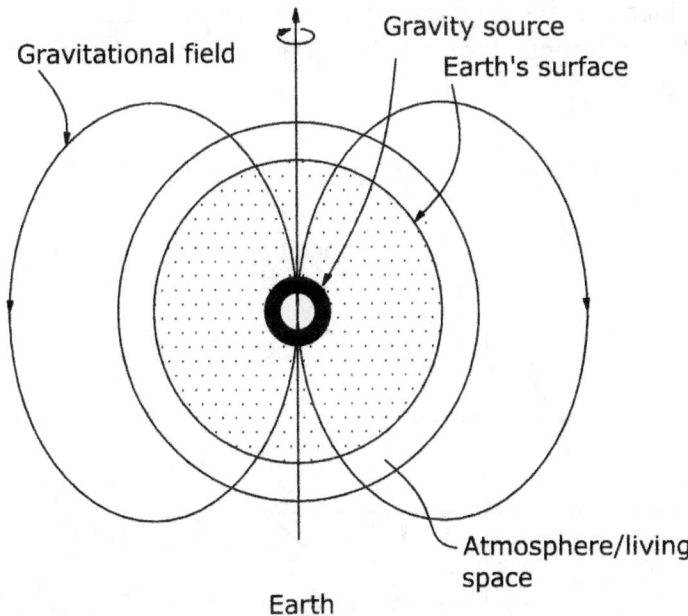

Fig. B.5. Parts of our planet. (The spin shown is the spin of the photon of the gravitational field of the Earth.)

Taking the structure of the Earth into account, we can design or deduce the parts of a flying saucer similar to a habitable gravitational producing body (Figure B.6).

* * *

Questions arise on what kind of energy is used to counteract gravity: whether it is electromagnetism, manmade gravitational field, the amount of energy of the photon emitted, or the volume of photon emitted.

Note that a flying saucer may be built without ever achieving the understanding of gravity. On the other hand, we may not be able to build an exact replica of a flying saucer based on the limitations of our materials or technology. For example, we are not a "recipient" of the product of a supernova that could

THEORY OF GRAVITY

have produced heavier elements than what is in our periodic table of the elements that are better source of "fuel" for our flying saucer.

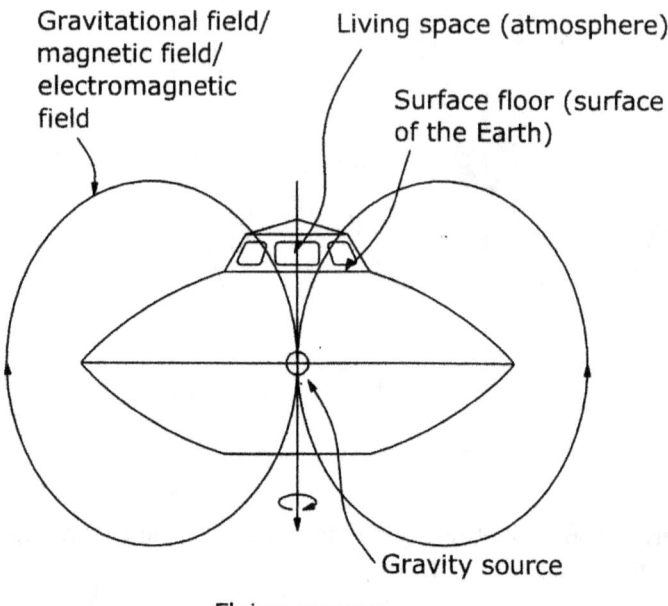

Fig. B.6. Parts of the flying saucer. (The spin shown is the spin of the photon of the gravitational field of the vehicle.)

Appendix C
Overthrowing Charge and Parity

The term charge parity is meaningful in the understanding of charge parity symmetry and charge parity violation. The term charge in charge parity refers to charge conjugation, which transforms (through mathematical operation) a particle into an antiparticle (for example, by changing the sign of the electric charge). The term parity in charge parity is called mirror symmetry (also called parity conservation) as it creates the mirror image of a particle. In mirror symmetry, it asserts that the laws of physics are indistinguishable between the events in the real world and those in the mirror.

CP symmetry postulates that the laws of physics should be the same if a particle is replaced (interchanged) with its anti-particle (C symmetry) and when its spatial coordinates are inverted (P symmetry or mirror symmetry). Thus, CP violation is the violation of the CP symmetry.

In 1964, James Cronin (born 1931) and Val Fitch (1923-2015) discovered CP violation in the decays of neutral kaons that resulted in their award of the Nobel Prize in Physics in 1980 "for the discovery of violations of fundamental symmetry principles in the decay of neutral K-mesons."

CP violation is said to play an important role in cosmology based on the Big Bang theory to explain the dominance of matter over antimatter in the present universe and in the study of weak interactions in particle physics. (Since the Big Bang never happened and that the question of the dominance of matter over antimatter had been settled in Chapter 9 on section *The Question of the Dominance of Matter Over Antimatter in the Universe: Big Bang Theory*, which I had argued that there is no such thing as antimatter, then the merit of the idea of CP violation is only if it is even right.)

* * *

I found the best explanation of charge (charge conjugation) and parity (mirror symmetry) in Leon Lederman's book *God Particle: If the Universe is the Answer, What is the Question?* on page 256. The book uses a spinning cylinder to illustrate charge (Figure C.1a) and charge parity violation (Figure C.1b).

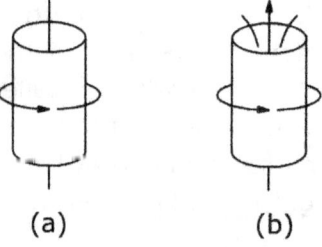

(a)   (b)

Fig. C.1. Illustration using a spinning cylinder to explain (a) charge and (b) charge parity violation of a decaying muon.

For my own purpose of explaining the charge and charge parity, I revised the book's explanation by replacing the cylinder with a particle (Figure C.2). Figure C.1a could easily be replaced with the particle shown in Figure C.2a. Figure C.1b, which is more of an emitted particle (free electron, free proton, photon) with its described spin, could be replaced with a particle with its spin and its direction of emission. (For the bound particle such as the fundamental particles within the proton and neutron, and the subatomic particles, the direction

Appendix C: Overthrowing Charge and Parity

of emission of its energy field is the same as its direction of emission as a free particle, which "preserves" its charge.)

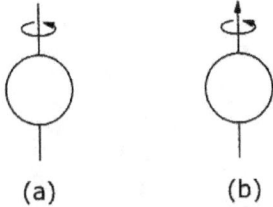

(a)　　　　　(b)

Fig. C.2. The illustration in Figure C.1 replaced with a spinning particle that (a) will illustrate the charge and (b) will illustrate the charge parity violation.

The book explained charge and parity (mirror symmetry) by explaining the spin and orientation of the particle (Figure 3).

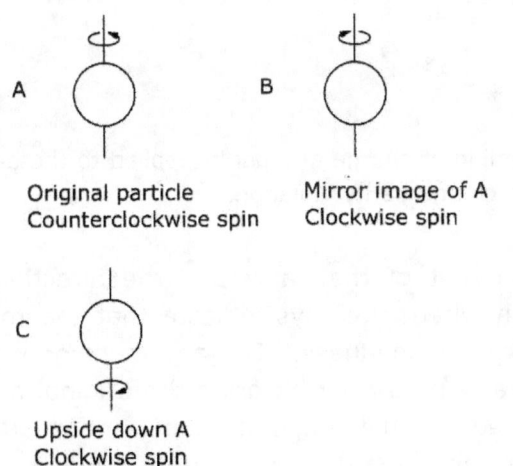

A　Original particle
Counterclockwise spin

B　Mirror image of A
Clockwise spin

C　Upside down A
Clockwise spin

Fig. C.3. Explanation of the charge and parity. Particle A is the original particle and particle B is the mirror image of particle A. Particle C is the upside down (understood as to give an opposite charge) of particle A, that is, by turning upside down particle A, its charge is understood to change.

Applying the understanding of charge and parity based on Figure C.3, this understanding is applied to charge parity violation using the decay of muon (Figure C.4).

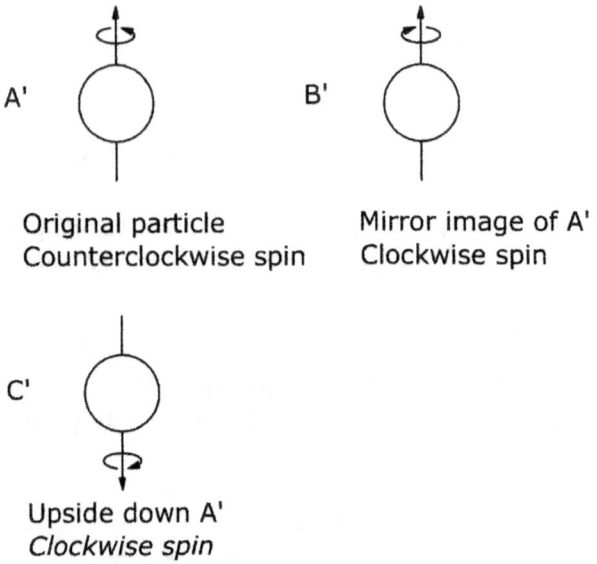

Fig. C.4. The understanding of charge and parity applied to the decay of muon to illustrate the charge parity violation.

The arrow coming out of the particle is the direction of emission of the muon where the rays indicate that the muon strongly preferred "right-handedness." The argument goes like this: Applying the idea of parity by mirroring the original muon into particle B' showed a left-handed decaying muon. If by experiment shows that all muon decays are right-handed, then particle B' does not exist in nature. Turning particle A' upside down into particle C' does not replicate particle B'. It is said that parity (mirror symmetry) is violated.

Appendix C: Overthrowing Charge and Parity

Refuting the Understanding of Charge Parity Symmetry and Charge Parity Violation

First and foremost, I have to point out that there is something wrong with how the particles in Figure C.1 are illustrated. Figure C.1a is the wrong illustration (and understanding) of charge. As I had discussed in this book, there are two types of particles: free particle (examples: free photon, free electron, and free proton) and bound particle (up quark, down quark, electron, proton, and neutron). Free particles travel in space (even the free electron that travels within the conductor or outside of the atom) while bound particles are within the atom. In Chapter 2 on section *Charge Theory: Energy Particle, Spin, and Energy Field*, I had discussed the direction of emission of the free particle identifies its charge. Free particle with a negative charge when viewed in front spins in a clockwise direction. (Bound particle with a negative charge when viewed from the top spins in a clockwise direction.) The direction of the spin of the negative charge particle agrees with the left-hand rule, which is used to determine the direction of the magnetic field when the direction of the flow of electron is known. For this reason, the illustration of the spin of muon (a family of electron) from Lederman's book, which is counterclockwise, is wrong as it should be clockwise.

The whole idea of charge and parity, and charge parity violation was founded on the wrong understanding of charge. To illustrate this, Figure C.2a is corrected to show the right illustration of a charge particle, which in this case is a free particle like that in Figure C.2b but spinning in a clockwise direction to show it with a negative charge.

Figure C.5 illustrates the wrong understanding of charge and parity that was shown in Figure C.3. (The charge of the original particle A is corrected to show that of a negative charge particle.)

THEORY OF GRAVITY

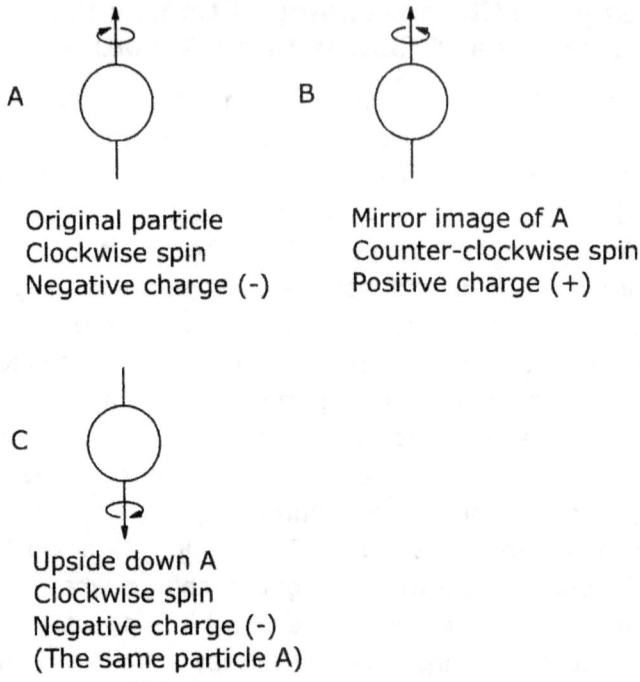

Fig. C.5. This figure illustrates that the understanding of charge and charge parity violation by the physicists had been wrong.

If particle A, which has a negative charge is "mirrored" as particle B, either particle B is an antiparticle of particle A or it is another opposite charge particle. For example, particle A is a free electron and particle B is (understood as a free particle) a positron. Positron is not the normal state of an electron, which means that it is either a product of a nuclear reaction or from a collision. The mirror image of a particle is its opposite charge or an illustration of a different particle with an opposite charge.

Particle C, which is the particle A that was turned upside down does not "mathematically" have an opposite charge to that of particle A. This is clearly demonstrated by a wire conductor with an electrical current positioned vertically. Turning the wire conductor upside down does not change the

Appendix C: Overthrowing Charge and Parity

charge of the electron. (In essence, parity is closer to the "mathematical" changing of the charge of a particle but purely on mathematical operation and has no bearing in the actual reality.)

So, comparing particle A to particle B is the same as comparing particle C to particle B. That is, the idea of mathematically turning upside down a particle to create its opposite charge is wrong. (The idea of turning upside a particle becomes nonsensical when applied to bound particles as the subatomic particles and the quarks do not have an upside down orientation. They cannot have an opposite charge within the atom as it is their innate charge within the atom to produce the matter of our universe.) Likewise, the idea of comparing the "turned upside down" particle's charge and mirror of the original particle to indicate the idea of charge parity violation does not make sense for reason that charge and parity and charge parity violation were a wrong mental construct of the physicists. That is, if nature had given the fundamental particles their charge, how could nature violate itself?

Notes and References

Chapter 1

1   Isaac Newton <http://en.wikipedia.org/wiki/Isaac_Newton> (9/30/2009) Note that the date of my source says September 2009. I had then finished my first book on a different field and started writing my first (unpublished) book in physics on my theory on light that I finished around 2011. I followed next with my book *Theory of Everything* that I self-published on September 2013. I started visiting my early outline for this book in the early 2014 and revising my outline in February or March.
2   Ibid.
3   Walter Isaacson. *Einstein: His Life and Universe* (New York: Simon and Schuster, 2007), 318.

Chapter 2

1   Inertia <http://en.wikipedia.org/wiki/Inertia> (3/2/2014)
2   Harris Bensons. *University Physics* (Canada: John Wiley & Sons, Inc., 1991), 69.
3   Ibid., 70.
4   Isaac Newton <http://en.wikipedia.org/wiki/Isaac_Newton> (9/30/2009).
5   Gravitational acceleration <http://en.wikipedia.org/wiki/Gravitational_acceleration> (3/19/14)
6   Galileo Galilei. *Dialogue Concerning the Two Chief World Systems* translated by Stillman Drake and edited by Stephen

Jay Gould (New York: Modern Library Paperback Edition, 2001), 164.

Chapter 3

1   Earth's core
<http://simple.wikipedia.org/wiki/Earth's_core> (12/09/2014)

Chapter 5

1   Expanding universe
<http://www.aip.org/history/cosmology/ideas/expanding.htm> (10/29/13)
2   Georges Lemaitre
<http://en.wikipedia.org/wiki/Georges_Lemaitre> (1/20/14)
A different version of the narrative was that in 1930, Eddington published in the *Monthly Notices of the Royal Astronomical Society* a long commentary on Lemaitre's 1927 paper, which Eddington described the work as a brilliant solution to the problem of cosmology. Lemaitre was invited to London where he proposed that the universe expanded from an initial point, which he called the "primeval atom." Lemaitre described his theory as "the Cosmic Egg exploding at the moment of creation."
3   The idea that can be derived from this is that neutron had existed since the Big Bang. Neutron is the result of nuclear fusion from the star, especially that of the hydrogen atom, which is composed of proton and electron. How could the early universe have neutrons if there was no nuclear fusion?
4   The problem with this argument is that radiation (heat) is created by nuclear fusion, radioactivity (changing of neutron to proton), or chemical reaction. That is, heat is a result of matter, it is either emitted or absorbed but it can never turn into particles of matter.

Chapter 8

1 Isaacson, 114.
2 Ibid., 116.
3 Ibid., 107.
4 Ibid., 131.
5 Ibid., 318.
6 Ibid., 189.
7 Ibid., 200.
8 Galileo Galilei. *Dialogue Concerning the Two Chief World Systems*. Translated by Stillman Drake and edited by Stephen Jay Gould (New York: Modern Library Paperback Edition, 2001), 164.
9 Timeline of luminiferous Aether <http://en.wikipedia.org/wiki/Timeline_of_lumiferous_aether> (1/24/2012)
10 Isaacson, 318.

Chapter 9

1 Peter Woit. *Not Even Wrong: The Failure of String Theory and the Search for Unity in Physical Law*. (New York: Basic Books, 2006), 70.
2 Soshichi Uchii. "Higgs Field," April 28, 2005 <http://www1.kcn.ne.jp/~h-uchii/Leib-Clk/higgs.html> (4/8/15)
3 Michael D. Lemonick, "Hunting the Higgs: Physicists May Soon Nab the Most Elusive Particle of All," *Time*, March 26, 2012, 16.
4 Experiment Brings Precision to a Cornerstone of Particle Physics <http://phys.org/news/2015-02-precision-cornerstone-particle-physics.html> (2/11/2015)

Bibliography

Aduana, Efren Jr. B. *Theory of Everything*. Publisher: Author (CreateSpace), 2013.

Benson, Harris. *University Physics*. Canada: John Wiley & Sons, Inc., 1991.

Descartes, René. *Principles of Philosophy*. Dordrecht, Holland: D. Reidel Publishing Company, 1984.

Fraser, Gordon, Egil Lillestøl, and Inge Sellevåg. *The Search for Infinity: Solving the Mysteries of the Universe*. New York: Facts On File, Inc., 1995.

Galilei, Galileo. *Dialogue Concerning the Two Chief World Systems*. Translated by Stillman Drake and edited by Stephen Jay Gould. New York: Modern Library Paperback Edition, 2001.

Isaacson, Walter. *Einstein: His Life and Universe*. New York: Simon and Schuster, 2007.

Lederman, Leon with Dick Teresi. *The God Particle: If Universe is the Answer, What is the Question?* New York: Mariner Books, 2006.

Moore, Patrick. *The Story of Astronomy.* new ed. New York: Grosset & Dunlap, 1977.
Previously published as "The Picture History of Astronomy"

Newton, Isaac and I. Bernard Cohen. *The Principia: Mathematical Principles of Natural Philosophy.* Berkeley and Los Angeles: University of California Press, 1999.

Woit, Peter. *Not Even Wrong: The Failure of String Theory and the Search for Unity in Physical Law.* New York: Basic Books, 2006.

Other References:

Carroll, Sean. *The Higgs Boson and Beyond.* The Great Courses, 2015. 2 Discs, 1 Book.

Singh, Simon. *Big Bang: The Origin of the Universe.* New York: Harper Perennial, 2005.

Smolin, Lee. *Three Roads to Quantum Gravity.* New York: Basic Books, 2001.

Westfall, Richard S. *The Life of Isaac Newton.* Cambridge: Cambridge University Press, 1993.

References Contradicting the Current Theories:

Arp, Halton. *Seeing Red: Redshifts, Cosmology and Academic Science.* Montreal, Canada: Apeiron, 1998.
(Critic of the Big Bang theory. Disputes the observation of an expanding universe.)

Baggott, Jim. *Farewell to Reality: How Modern Physics Has Betrayed the Search for Scientific Truth.* New York, New York: Pegasus Books, 2013.

Bibliograpy

Lerner, Eric J. *The Big Bang Never Happened*. New York: First Vintage Books Edition, 1992.
(Espoused the plasma theory. Critic of the Big Bang theory.)

Smolin, Lee. *The Trouble with Physics*. New York: First Mariner Books, 2007.
(Critic of the string theory.)

Unzicker, Alexander. *The Higgs Fake: How the Particle Physicists Fooled the Nobel Committee*. Publisher: Author (CreateSpace), 2013.
(Critic of how the LHC had discovered the Higgs boson.)

Woit, Peter. *Not Even Wrong: The Failure of String Theory and the Search for Unity in Physical Law*. New York: Basic Books, 2006.
(Critic of the string theory.)

Acknowledgements

I would like to thank my mother Rachel and father Efren Sr. for their financial support all these years. I would like to thank my brother Gleen and his wife Joy for their support. I would like to thank them for the understanding and patience that they gave me in the path that I had chosen.

About The Author

Efren Basa Aduana Jr. is a graduate of Bachelor of Science in Electrical Engineering in the Philippines. He is an inventor and a writer. He had patented two simple inventions and had been writing full time since the middle of 2006.[3.5]

ebaduanajr@gmail.com

www.ingramcontent.com/pod-product-compliance
Lightning Source LLC
Chambersburg PA
CBHW070225190526
45169CB00001B/79